新编全国高等职业院校烹饪专业规划教材

烹饪英语

CUISINE ENGLISH

华路宏◎主编

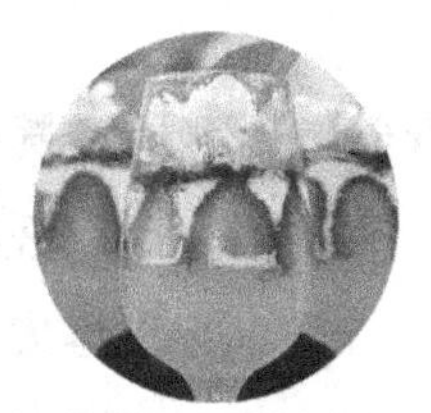

北京·旅游教育出版社

责任编辑：果凤双

图书在版编目（CIP）数据

烹饪英语 / 华路宏主编. -- 北京 ：旅游教育出版社，2017.5（2024.7重印）
新编全国高等职业院校烹饪专业规划教材
ISBN 978-7-5637-3561-7

Ⅰ. ①烹… Ⅱ. ①华… Ⅲ. ①烹饪—英语—高等职业教育—教材 Ⅳ. ①TS972.1

中国版本图书馆CIP数据核字(2017)第090380号

新编全国高等职业院校烹饪专业规划教材
烹饪英语
华路宏　主编

出版单位	旅游教育出版社
地　　址	北京市朝阳区定福庄南里1号
邮　　编	100024
发行电话	（010）65778403　65728372　65767462（传真）
本社网址	www.tepcb.com
E - mail	tepfx@163.com
排版单位	北京旅教文化传播有限公司
印刷单位	北京虎彩文化传播有限公司
经销单位	新华书店
开　　本	710毫米×1000毫米　1/16
印　　张	6.75
字　　数	85千字
版　　次	2017年5月第1版
印　　次	2024年7月第4次印刷
定　　价	22.00元

（图书如有装订差错请与发行部联系）

出版说明

我国烹饪享誉世界。进入21世纪以来,随着社会经济的发展和人们生活水平的不断提高,国际化交流不断深入,烹饪行业经历了面临机遇与挑战、兼顾传承与创新的巨大变革。烹饪专业教育教学结构也随之发生了诸多变化,我国烹饪教育已进入了一个蓬勃发展的全新阶段。因此,编写一套全新的、能够适应现代职业教育发展的烹饪专业系列教材,显得尤为重要。

本套“新编全国高等职业院校烹饪专业规划教材”是我社邀请众多业内专家、学者,依据《国务院关于加快发展现代职业教育的决定》的精神,以职业标准和岗位需求为导向,立足于高等职业教育的课程设置,结合现代烹饪行业特点及其对人才的需要,精心编写的系列精品教材。

本套教材的特点有:

第一,推进教材内容与职业标准对接。根据职业教育“以技能为基础”的特点,紧紧把握职业教育特有的基础性、可操作性和实用性等特点,尽量把理论知识融入实践操作之中,注重知识、能力、素质互相渗透,契合现代职业教育体系的要求。

第二,以体现规范为原则。根据教育部制定的高等职业教育专业教学标准及劳动和社会保障部颁布的执业技能鉴定标准,对每本教材的课程性质、适用范围、教学目标等进行规范,使其更具有教学指导性和行业规范性。

第三,确保教材的权威性。本套教材的作者均是既具有丰富的教学经验又具有丰富的餐饮、烹饪工作实践经验的专家,熟悉烹饪专业教学改革和发展情况,对相关课程的教学和发展具有独到见解,能将教材中的理论知识与实践中的技能运用很好地统一起来。

第四,充分体现教材的先进性和前瞻性。在现代科技发展日新月异的大环境下,尽量反映烹饪行业中的新工艺、新理念、新设备等内容,适当展示、介绍本学科最新研究成果和国内外先进经验,体现教材的时代特色。

第五,体例新颖,结构科学。根据各门课程的特点和需要,结合高等职业教育规范以及高职学生的认知能力设计体例与结构框架,对实操性强的科目进行模块化构架。教材便于学生开阔视野,提升实践能力。

作为全国唯一的旅游教育专业出版社,我们有责任也有义务把体现最新教学改革精神、具有普遍适用性的烹饪专业教材奉献给大家。在这套精心打造的教材即将面世之际,深切地希望广大教师学生能一如既往地支持我们,及时反馈宝贵意见和建议。

旅游教育出版社

前　言

《烹饪英语》是作者多年来在烹饪英语教学研究、课程改革、课堂实践的基础上，汲取作者主编的《厨房情景英语》的精华，调研杭州各大餐饮企业之后而编写的烹饪专业英语实用教材。

本书专门针对高等职业技术院校烹饪专业的英语教学，同时也适用于各大涉外宾馆、酒店餐饮行业员工培训及相关烹饪专业人员自学。本书主要内容包括：厨房人事结构、厨师操作注意事项、厨房设备用具、烹饪原料、菜肴制作过程及西餐礼仪等。

《烹饪英语》以理论与实践相结合，注重烹饪英语语言的沟通交流实践作为编写理念，力求提高烹饪专业学生和烹饪行业从业人员的烹饪英语沟通交流能力；并为学员进一步运用英语语言工具，研究国内外烹饪行业最新发展动态，跟踪国内外烹饪行业研究成果打好英语语言基础，因而也带有部分工具书的功能。

本书紧扣烹饪行业英语交流实际，图文并茂，深入浅出，内容全面。课堂教学中，建议教师不单纯讲解，而是把学生作为教学主体，引导学生参与整个教学过程，实现师生之间、学生之间充分的互动，从而培养学生的独立思考能力和动口能力。

《烹饪英语》由浙江旅游职业学院教师华路宏主持编写。在编写过程中，编者多次与杭州西溪喜来登酒店、浙旅集团机场大酒店等餐饮企业行政总厨沟通，获取了丰富翔实的第一手资料；浙江旅游职业学院烹饪系对本书提供了资料，并提出修改意见，做出具体指导并提供其他各方面的支持；编写过程中烹饪系学生协助整理资料；旅游教育出版社为本书的出版付出了辛勤的劳动；华恭武先生和应英女士为本书原料图片拍摄提供了帮助。在此谨表示衷心感谢！

本书的编写，受时间仓促、编者水平的局限，存在不足之处，期待业内专家学者、使用本书的师生以及广大读者提出宝贵意见，以供进一步完善本书之用。

编者

2016 年 12 月 29 日

前　言

Contents
目录

Unit 1

Kitchen Titles and Rules

Lead in

A chef is a highly trained and skilled professional cook who is proficient in all aspects of food preparation of a particular cuisine. The word "chef" is derived (and shortened) from the term *chef de cuisine*, the director or head of a kitchen. Chefs can receive both formal training from an institution, as well as through apprenticeship with an experienced chef.

Vocabulary

cook [kʊk] *n.* 厨师
chef [ʃef] *n.* 厨师；主厨
executive [ɪg'zekjətɪv] *adj.* & *n.* 执行的；决策者
pastry ['peɪstri] *n.* 面点
assistant [ə'sɪstənt] *adj.* & *n.* 助理的；助理
vegetable ['vedʒtəbl] *n.* 蔬菜；植物
butcher['bʊtʃə(r)] *n.&vt.* 屠夫；屠宰
sauce [sɔːs] *n.&vt.* 酱汁；给……调味
soup[suːp] *n.* 汤；羹
grill[grɪl] *n.&vt.* 烤架；在（烤架上）烤
roast [rəʊst] *n.&vt.* 烤肉；烘烤
staff [stɑːf] *n.* 全体人员
aboyeur [abwajuː] *n.* 跑堂喊菜的人

commis ['kɒmi] *n.*（法语）小职员；副手
apprentice [ə'prentɪs] *n.* 学徒；见习生
clerk[klɑːk] *n.* 职员；办事员
relief[rɪ'liːf] *n.* 替代
pantryman['pæntrɪmən] *n.* 配餐员；司膳总管；司膳总管助理
potman['pɒtmən] *n.* 酒馆的侍者
porter['pɔːtə(r)] *n.* 搬运工；（列车）服务员；杂务工
steward['stjuːəd] *n.* 乘务员；管家；干事；管理员

List of Key Words

（1）Executive chef 行政总厨师长
（2）Assistant chef 行政总厨师长助理
（3）Sous-chef 副厨师长，指具体负责并干活的厨师长
（4）Larder chef 负责烹饪各种肉类的厨师长
（5）Pastry chef 负责烹饪各种面点的厨师长
（6）Vegetable chef 负责烹饪各种蔬菜的厨师长
（7）Butcher 屠夫，负责屠宰各种禽类的人
（8）Grill cook 负责在烤架上烤炙肉类的厨师
（9）Sauce cook 负责调制酱汁的厨师
（10）Soup cook 负责烹饪各种汤的厨师
（11）Fish cook 负责烹饪鱼类的厨师
（12）Roast cook 负责在烤箱内烧烤肉类的厨师
（13）Breakfast cook 负责烹饪早餐的厨师
（14）Night cook 晚上上班的厨师
（15）Staff cook 负责烹饪员工伙食的厨师
（16）Relief cook 替班厨师，后补厨师。此人是个多面手，任何部门的厨师因休假或病假不能上班，替班厨师就补上
（17）Commis 助理厨师，是主厨的副手
（18）Aboyeur 跑堂喊菜的服务员，他忙于餐厅和厨房之间，把客人的点菜单送给厨师们
（19）Kitchen clerk 厨房会计，负责厨房的一切文书工作
（20）Pantryman 负责管理配膳室（食品小贮藏室），不是厨师，不烹饪
（21）Potman 负责擦洗大深锅的人

Focus on Language

Dialogue 1

Susan(S): Excuse me, where is the executive chef's office?

Commis(C): Go straight and turn right at the end of the hallway. Then it's on your left hand side.

S: Thank you. I just want to ask him about the salary.

C: OK. But maybe he is in the beverage cooler now.

S: How long will he be back or shall I go to the beverage cooler to find him?

C: He will come back in 20 minutes. You can wait in his office. I can guide you.

S: Oh, thank you so much!

C: My pleasure.

Dialogue 2

Sous chef (S): Nice to meet you. I am Kingsley King, the sous chef working in Feast Restaurant. Welcome to join us.

Commis chef (C): Nice to meet you, too, Mr. King. My name is Adam Green. What should I do as a commis cook?

S: You should first learn how to help other cooks while they are working. I'm in charge of cooking and staff training. And I will tell you how to do it.

C: Thank you so much. You are so approachable. I will try my best to do it. By the way, who is the man over there?

S: Oh, it's Peter Wong, the larder chef in our restaurant.

C: Larder chef? What does he do?

S: He is in charge of cooking all kinds of meat.

C: Oh, I see. Thanks.

Task 1 Do you know the names of kitchen positions? Please look at the graph and try to fill in the form.

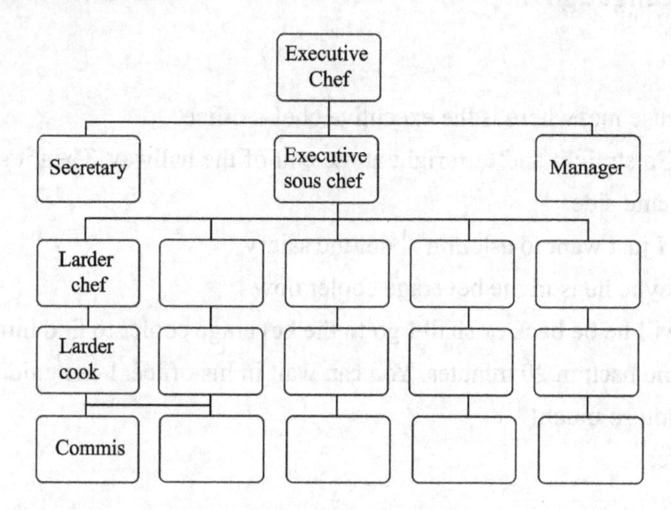

Task 2 Can you identify different kinds of kitchen position? Write down the titles and match the exact positions with them.

(1) He is the head of chefs and works in the chef's office. ________

(2) He is in charge of all cooking things. ________

(3) He is in charge of cooking all kinds of meat. ________

(4) He helps other cooks. ________

(5) He is in the hot kitchen to roast meat. ________

(6) He is in the scullery. ________

(7) He is in the vegetable preparation section. ________

(8) He is in the hot kitchen to grill meat. ________

(9) He works in the kitchen store. ________

(10) He works in the fish section. ________

(11) He works in the pastry section. ________

(12) He works in the butchery. ________

Task 3 Look at the pictures and try to write down the title of those people below the pictures.

Task 4 Practise the useful sentences below.

(1) What does he/she do?

What do you do?

What does your father/mother/sister do?

(2) I'm a sous chef/commis cook working in...

(3) I am in charge of...

I am responsible for...

(4) I am very glad to introduce myself to you.

(5) He is in the hot kitchen to...

(6) I just want to ask him/her about...

Task 5 Role play.

Setting: 招聘行政总厨

Work in groups, role play in the setting of taking job application of executive chef. The scene set as below:

- ▶ up to 22 000 pounds per year
- ▶ 28 days holidays + benefits
- ▶ Bridge pub, west London

Task 6 Translate the following sentences into English.

1. 请问你在厨房里是做什么的?

____

2. 请问学徒一天在厨房里要工作多久?

____

3. 蔬菜厨师主要负责烹饪各类蔬菜菜肴。

____

4. 你是如何找到洗碗工这份工作的?

____

5. 布莱恩是一名屠夫,他每天要在厨房里屠宰各种家禽。

____

Reading

Titles

Various titles, detailed below, are given to those working in a professional kitchen and each can be considered a title for a type of chef. Many of the titles are based on the *brigade de cuisine* (or brigade system) documented by Auguste Escoffier, while others have a more general meaning depending on the individual kitchen.

Chef de cuisine

Other names include executive chef, chef manager, head chef, and master chef. This person is in charge of all activities related to the kitchen, which usually includes menu creation, management of kitchen staff, ordering and purchasing of inventory, and plating design. *Chef de cuisine* is the traditional French term from which the English word chef is derived. *Head chef* is often used to designate someone with the same duties as an executive chef, but there is usually someone in charge of a head chef, possibly making the larger executive decisions such as direction of menu, final authority in staff management decisions, and so on. This is often the case for executive chefs with multiple restaurants. Involved in checking the sensory evaluation of dishes after preparation and they are well aware of each sensory property of those specific dishes.

Sous-chef

The *sous-chef de cuisine* (under-chef of the kitchen) is the second-in-command and direct assistant of the Chef de Cuisine. This person may be responsible for scheduling the kitchen staff, or substituting when the head chef is off-duty. Also, he or she will fill in for or assist the *Chef de partie* (line cook) when needed. This person is accountable for the kitchen's inventory, cleanliness, organization, and the continuing training of its entire staff. A sous-chef's duties can also include carrying out the head chef's directives, conducting line checks, and overseeing the timely rotation of all food products. Smaller operations may not have a sous-chef, while larger operations may have more than one. The sous chef is also responsible when the executive chef is absent.

Chef de partie

A *chef de partie*, also known as a "station chef" or "line cook", is in charge of a particular area of production. In large kitchens, each chef de partie might have several cooks or assistants. In most kitchens, however, the chef de partie is the only worker in that department. Line cooks are often divided into a hierarchy of their own, starting with "first cook", then "second cook", and so on as needed.

Commis (Chef) / Range chef

A *commis* is a basic chef in larger kitchens who works under a *chef de partie* to learn the station's or range's responsibilities and operation. This may be a chef who has recently completed formal culinary training or is still undergoing training.

Unit 2

Kitchen Safety and Kitchen Hygiene

Lead in

Culinary hygiene pertains to the practices related to food management and cooking to prevent food contamination, prevent food poisoning and minimize the transmission of disease to other foods, humans or animals. Culinary hygiene practices specify safe ways to handle, store, prepare, serve and eat food.

Vocabulary

hygiene ['haɪdʒiːn] n. 卫生；卫生学
uniform ['juːnɪfɔːm] n. 制服，操作服
storage [ˈstɔːridʒ] n. 存储
rinse [rɪns] n. 冲洗
butchery ['bʊtʃəri] n. 屠杀；屠场；屠宰业
rodent ['rəʊdnt] n. 啮齿动物
insect ['ɪnsekt] n. 昆虫；虫子
rat [ræt] n.&v. 变节者；老鼠；捕鼠
bacteria [bæk'tɪəriə] n. (复数)细菌
cockroach ['kɒkrəʊtʃ] n. 蟑螂
poison ['pɔɪzn] n.&vt.&adj. 毒药；毒害；有毒的；危险的
sanitize ['sænɪtaɪz] vt. 采取卫生措施使其安全；消毒
antiseptic [ˌænti'septɪk] n.&adj. 杀菌剂；防腐剂；杀菌的；防腐的
detergent [dɪ'tɜːdʒənt] n.&adj. 清洁剂；用于清洗的

extinguisher[ɪkˈstɪŋgwɪʃə] *n.* 灭火器
washer section 锅具清洗区
beverage cooler 饮料冷库
kitchen store 厨房贮藏室
pick up area 备菜间
scullery section 餐具洗涤区

Focus on Language

Dialogue 1

Apprentice(A): I am going to work in this kitchen. what rules shall I know then?

Chef(C): You should pay attention on the kitchen hygiene and kitchen safety.

A: Could you tell me some details about kitchen hygiene and kitchen safety?

C: Certainly. There may be rats and cockroaches in the kitchen for the reason that there are full of food there.

A: What shall we do?

C: Don't panic, just poison them or trap them and everything in the kitchen should be sanitized. And also the bacteria may get in food.

A: How do microbes or bacteria get in food?

C: Usually microbes get into the food from the hands of people working in the kitchen. So we should always wash our hands.

A: Get it. What kind's of foods are most dangerous?

C: Protein foods: like meat, poultry, eggs, fish, diary products, etc.

A: How can we stop food poisoning?

C: Never leave food outside the refrigerator for more than two hours and keep refrigerator food below 40 ℉.

Dialogue 2

A: Chef, what is the most dangerous thing in the kitchen?

C: Safety is the number one priority in the kitchen. Let me show you some dangerous things and the methods we could avoid in the kitchen.

A: Yes, I will listen carefully.

C: Firstly, we should pay attention on the "fire". Oily rags are dangerous. Wash them in soapy water or throw them away, using deep fryers are dangerous. Watch out for smoke, foam, bubbles. You should learn how to put out an oil or grease fire, you need to have carefully supervised practice in putting out oil and grease fires with a proper type of chemical fire extinguisher. Do not use water on an electrical fire!

A: I think I have to be taught the knowledge about it.

C: Secondly, you should pay attention on "burning hurt". Hot oil and melted fat can easily burn you. If you are not careful, cooking with grease and oil is quite dangerous. They can quickly start a big kitchen fire. By the way, damaged electrical wires are also very dangerous. Do not use the damaged electrical equipment.

A: Oh, really? I am afraid of being burned, I'll be careful.

C: Last but not least, you need avoid slipping or tripping. Falling down can be quite dangerous. If you are carrying something hot, you can also get badly burned; if you are carrying something made out of glass or something sharp, you can also get a deep cut because of slippery.

A: I think I'd better be more careful. Thank you so much!

Task 1 Look at the picture below, the kitchen is clean and safe. Fill in the blankets in the sentences with the verbs or phrases given if you want to keep the kitchen like the picture.

wipe up　　put out　　sanitize　　wash　　clean

1. You should learn how to ________ fire with extinguisher.
2. Pick up everything you drop, ________ everything you spill.
3. I should always ________ my hands.
4. We have to ________ all the things before we leave the kitchen.

5. Everything in the kitchen should be ________ .

Task 2 The followings are some kitchen rules you should know. Could you add any other rules as many as possible?

☞ Kitchen supplies cannot be taken home.

☞ Kitchen doors cannot be propped open but the doors on the North and South sides can be lifted for ventilation purposes.

☞ Kitchen utensils cannot be borrowed.

☞ Proper food preparation cannot be postponed.

☞ Everyone is not allowed to wear shorts in the restaurant.

☞ Sinks must be cleaned after use.

☞ Dish sink and surrounding areas should be cleaned and wiped dry after use.

☞ Clean clothing and close-toed shoes must be worn while working in the kitchen.

Task 3 Read the following numbered English words about Kitchen Floor and try to write out their Chinese versions.

<table>
<tr><td>1. Freezer</td><td>2.Cold Kitchen</td><td>3.Butchery</td><td>4.Pastry</td><td>5. Beverage Cooler</td><td rowspan="3">13. Pick-up Area</td></tr>
<tr><td rowspan="2">12.Kitchen Store/Pantry Area</td><td colspan="4">6. Chef's Office
7. Hot Kitchen</td></tr>
<tr><td>8. Pot-washer</td><td>9.Vegetable Preparation</td><td>10. Fish Section</td><td>11. Scullery</td></tr>
</table>

1. ______________________________
2. ______________________________
3. ______________________________
4. ______________________________
5. ______________________________
6. ______________________________
7. ______________________________
8. ______________________________

9. ____
10. ____
11. ____
12. ____
13. ____

Task 4 Translate the following sentences into English.

1. 进厨房工作之前请穿好操作服。

____

2. 烹饪前请先洗手。

____

3. 请把厨房地板的油污冲洗干净。

____

4. 处理食物前，保持双手干净，剪短指甲。

____

5. 你应该把蔬菜放到冰箱里储藏。

____

6. 厨房原料使用应该遵循“先进先出”的原则。

____

Reading

A kitchen is a room or part of a room used for cooking and food preparation in a dwelling or in a commercial establishment. In the West, a modern residential kitchen is typically equipped with a stove, a sink with hot and cold running water, a refrigerator, counters and kitchen cabinets arranged according to a modular design. Many households have a microwave oven, a dishwasher and other electric appliances. The main function of a kitchen is serving as a location for storing, cooking and preparing food (and doing related tasks such as dishwashing), but it may also be used for dining, entertaining and laundry.

Commercial kitchens are found in restaurants, cafeterias, hotels, hospitals, educational and workplace facilities, army barracks, and similar establishments. These kitchens are generally larger and equipped with bigger and more heavy-duty equipment than a residential kitchen. For example, a large restaurant may have a huge walk-in refrigerator and a large commercial dishwasher machine. Commercial kitchens are generally (in developed countries) subject to public health laws. They are inspected periodically by public-health officials, and forced to close if they do not meet hygienic requirements mandated by law.

Unit 3

Cooking Method and Cooking Verbs

Lead in

Cooking or cookery is the art, technology and craft of preparing food for consumption with the use of heat. Cooking techniques and ingredients vary widely across the world, from grilling food over an open fire to using electric stoves, and to baking in various types of ovens, reflecting unique environmental, economic, and cultural traditions and trends. The ways or types of cooking also depend on the skill and type of training an individual cook has. Cooking is done both by people in their own dwellings and by professional cooks and chefs in restaurants and other food establishments. Cooking can also occur through chemical reactions without the presence of heat, most notably with ceviche, a traditional South American dish where fish is cooked with the acids in lemon or lime juice.

Vocabulary

pastry ['peɪstri] *n.* 面粉糕饼；馅饼皮

dough [dəʊ] *n.* 生面团

twist [twɪst] *v.* 缠绕；捻；搓；拧；扭曲

fold [fəʊld] *vt.&vi.&n.* 折叠；交叉；折层；折痕

knead [ni:d] *v.* 揉(面粉等)；按摩；捏制

stretch [stretʃ] *v.&n.&adj.* 伸展；延伸；可伸缩的

flatten ['flætn] *vt.&vi.* 使变平

braid [breɪd] *v.* 编织

rub [rʌb]	*v.* 擦；揉；搓
wrap [ræp]	*v.* 包；裹；覆盖
prick [prɪk]	*v.* 刺；扎
garnish ['gɑːnɪʃ]	*v.* &*n.* 装饰；配菜
demonstrate ['demənstreɪt]	*vt.*&*vi.* 示范；演示
rinse [rɪns]	*n.*&*v.* 清洗；冲洗
scale [skeɪl]	*n.*&*v.* 刻度；等级；刮鳞；剥落
gut [gʌt]	*n.*&*vt.* 内脏；取出内脏
cut off	*vt.*&*vi.* 切断；停止运行
flake [fleɪk]	*n.*&*v.* 薄片；剥落
bone [bəʊn]	*n.*&*v.* 骨；除去骨头
lard [lɑːd]	*n.*& *v.* 猪油点缀；涂油于；润
wring out	*vt.* 绞出；拧掉；扭干
soak[səʊk]	*n.*&*v.* 浸；浸泡；渗透
cube [kjuːb]	*n.*&*v.*&*adj.* 立方；求……的立方；立方的
mince [mɪns]	*v.*&*n.* 切碎；肉酱
dice [daɪs]	*n.*&*v.* 骰子；小立方；把……切成丁
peel [piːl]	*n.*&*v.* 果皮；削皮
grate [greɪt]	*v.* 摩擦；磨碎
crush [krʌʃ]	*n.*&*v.* 压碎；压榨
squeeze [skwiːz]	*n.*&*v.* 挤压；塞进；压榨
grease[griːs]	*n.*&*vt.* 油脂；用油脂涂；上油
stuff[stʌf]	*n.*&*vt.* 塞满；填满
marinate ['mærɪneɪt]	*v.*（做色拉时将肉鱼等在调味汁中）浸泡；腌制
strain [streɪn]	*n.*&*v.* 拉紧；紧张
sift [sɪft]	*v.* 筛
skim [skɪm]	*n.*&*v.* 浮沫；撇去；掠过
saute ['səʊteɪ]	*n.*&*v.*&*adj.* 嫩煎菜肴；快炒；煎的；炒的
rolling pin	擀面杖
knock back	揉打
cut open	剖；切开
tie up	扎紧
trim off	修剪
mash up	捣烂

List of Key Expressions

处理面团工序　Knock back the pastry.
Twist the pastry.
Fold the pastry.
Braid the dough.
Knead the dough.
Rub the flour and butter together.
Roll out the pastry.
Flatten the pastry.
Shape the pastry.
Divide the dough.
Let's put a layer of chocolate icing on the cake.
Layer cream on the cake.
Garnish the cake with whole cherries.

处理鱼类工序　Rinse the fish.
Cut off the fins.
Scale it. Then gut it.
Cut open the stomach of the fish. Then take out the guts.

处理肉类工序　To bone the beef.
Tie up the meat.
To flatten the meat.
To beat the meat.

剖开　Open the coconut.

切块　Cut it into pieces.

切成条　Please cut the beef into strips.

切丝　Shred the cheese.

切片　Please slice the aubergines.

切成细末　Should I chop the onions fine?

切碎　① I need to cut up some onions.
② Mince the beef.

切丁　Should I dice the carrots?

切开　Should I split them down the middle?

切除 ① Should I trim off the ends?
② Cut up the bones.

剥（削）皮 ① Peel ten cloves of garlic.
② Peel an apple.

去籽 ① Please remove the seeds.
② Seed the pumpkin.

磨碎 Grate the cheese.

捣成泥 Mash up the bananas.

填充 Please stuff the peppers.

挤压 Squeeze the lemons.

打蛋 Beat the eggs.

沥干 To drain them.

过滤 Strain sauces.

去除表面杂质 Skim the liquid.

过筛 Sift the breadcrumbs.

刷上奶油 Brush the bread.

抹奶油 Butter the mushrooms.

加盐 Salt the cauliflower.

涂抹 Spread the jam.

腌渍 I want to marinate some chicken.

融化 Melt some butter.

预热 Preheat the oven to 350 degrees.

油炸 Fry the chicken.

蒸 Steam the rice.

煎 Fry the crepes.

嫩煎 Saute the peas.

调味 Season the soup before serving.

撒上 Sprinkle some cheese on the pizzas.

上菜 Serve the potatoes.

Focus on Language

Dialogue 1

Apprentice(A): Good morning, chef! What are we going to make today?

Chef(C): Morning, today we are going to make fruit cake.

A: What shall we do first?

C: Prepare the pastry first. Do you know how to work the pastry with a rolling pin?

A: Sorry, could you demonstrate once for me?

C: All right, now let's begin.

A: What kind of fruit do we need to add?

C: Some mango and blueberry.

A: What should I do next?

C: Slice the mango into half,remove the core,cube the mango and scoop out it.

A: OK, after I finish the cake, do I need to place the fruit on it?

C: Of course,with some cream.

Dialogue 2

Lisa(L): What kind of pasta do you want?

Sherry(S): We can try the spinach pasta. Maybe it's healthier.

L: I hope Nigel doesn't complain.

S: Don't worry. He's not finicky. We still need to get tomato sauce.

L: Right. Does he like spaghetti or…?

S: He likes spaghetti. he likes Chinese noodles, too.

L: I know a good brand: Paul Newman's. I hope they still sell it.

S: So where is the tomato sauce?

L: It's over there.

S: Let's see. Beef, onions, garlic, pasta, tomato sauce. What else do we need?

L: Lettuce for salad.

S: What kind of salad will you make?

L: Caesar salad.

◇◇

Task 1 Try to write down the English name or action below the pictures.

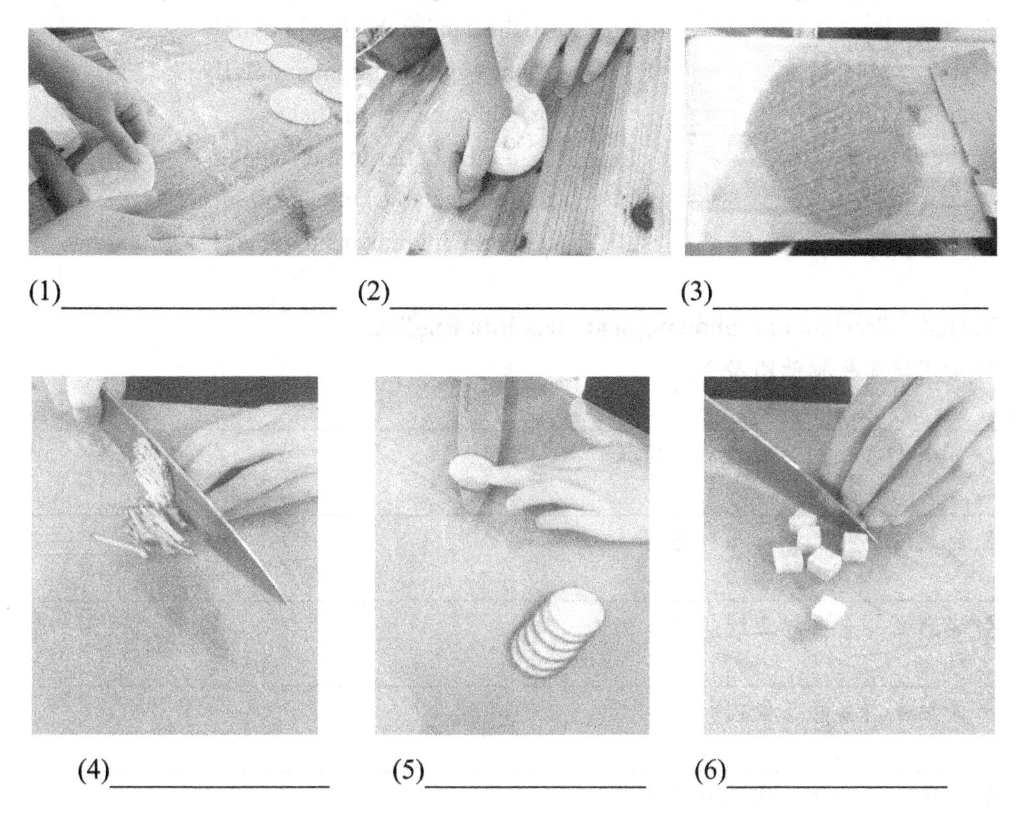

(1)__________________ (2)__________________ (3)__________________

(4)________________ (5)________________ (6)________________

Task 2 Put the following sentences into the correct order according to the time sequence.

1. Cook the fish.
2. Gut the fish.
3. Rinse the fish.
4. Scale the fish.
5. Pan-fry the fish.

________ ________ ________ ________ ________

Task 3 Match the verbs on the left with the phrases on the right.

1. squeeze	a. the mixture with a wooden spoon
2. melt	b. the potatoes and boil in a pan
3. beat	c. the cheese and add to the sauce

4. mix d. the sauce over the meat and serve
5. chop e. the ham as thinly as possible
6. stir f. the eggs until light and fluffy
7. grate g. a lemon over the fish
8. slice h. a little butter in a frying pan
9. pour i. the vegetables into small pieces
10. peel j. all the ingredients together

Task 4 Traslate the following sentences into English.

1. 请用擀面杖把面团擀平。

____

2. 那个小男孩把面团揉成了球状。

____

3. 我想给蛋糕上撒一层糖霜。

____

4. 烧鱼之前先去掉鱼鳞。

____

5. 请把牛肉去骨并把肉拍平。

____

6. 把鱼肚子切开，然后取出内脏。

____

7. 请把这个木瓜去籽。

____

8. 玛丽让丈夫帮她把烤箱预热到 180 度。

____

9. 服务员往我的饮料中挤了少量柠檬汁。

____

Reading

Doughs vary widely depending on ingredients, the kind of product being produced, the type of leavening agent (particularly whether the dough is based on yeast or not), how the dough is mixed (whether quickly mixed or kneaded and left to rise), and cooking or baking technique. There is no formal definition of what makes dough, though most doughs have viscoelastic properties.

Leavened or fermented doughs (generally made from grain cereals or legumes that are ground to produce flour, mixed with water and yeast) are used all over the world to make various breads. Salt, oils or fats, sugars or honey and sometimes milk or eggs are also common ingredients in bread dough. Commercial bread doughs may also include dough conditioners, a class of ingredients that aid in dough consistency and final product.

Unit 4
Vegetables

Lead in

Vegetables have been part of the human diet from time immemorial. Some are staple foods but most are accessory foodstuffs, adding variety to meals with their unique flavors and at the same time, adding nutrients necessary for health. Some vegetables are perennials, but most are annuals and biennials, usually harvested within a year of sowing or planting. Whatever system is used for growing crops, cultivation follows a similar pattern: preparation of the soil by loosening it, removing or burying weeds and adding organic manures or fertilisers; sowing seeds or planting young plants; tending the crop while it grows to reduce weed competition; controlling pests and provide sufficient water; harvesting the crop when it is ready; sorting, storing and marketing the crop or eating it fresh from the ground.

Vocabulary

asparagus[əˈspærəgəs] *n.* 芦笋
chive[tʃaiv] *n.* 香葱
scallion [ˈskæljən] *n.* 青葱
garlic [ˈgɑrlɪk] *n.* 大蒜
shallot [ʃəˈlɑt, ˈʃælət] *n.* 葱
onion [ˈʌnjən] *n.* 洋葱
water chestnut [ˈwɔtɚ ˈtʃɛsˌnʌt] *n.* 菱角，荸荠

turnip [ˈtə:nɪp] *n.* 萝卜
carrot [ˈkærət] *n.* 胡萝卜
radish [ˈrædɪʃ] *n.* 小萝卜
eggplant [ˈɛgˌplænt] *n.* 茄子
green pepper [grin ˈpɛpɚ] *n.* 青椒
olives [ˈɑlɪv] *n.* 橄榄
cucumber [ˈkjuˌkʌmbɚ] *n.* 黄瓜
okra [ˈokrə] *n.* 黄秋葵
tomato [təˈmeto] *n.* 西红柿
pumpkin [ˈpʌmpkɪn, ˈpʌm-, ˈpʌŋ-] *n.* 南瓜
spinach [ˈspɪnɪtʃ] *n.* 菠菜
lettuce [ˈlɛtəs] *n.* 莴苣
cabbage [ˈkæbɪdʒ] *n.* 甘蓝
celery [ˈsɛləri] *n.* 芹菜
potato [pəˈteto] *n.* 马铃薯
cauliflower [ˈkɔlɪˌflaʊɚ] *n.* 花椰菜
turnip [ˈtə:nɪp] *n.* 芜青，萝卜
purslane[ˈpɚslɪn, -ˌlen] *n.* 马齿苋
arugula[əˈrugələ] *n.* 芝麻菜
cress[krɛs] *n.* 水芹；水蔊菜
radicchio[rəˈdikio, rɑ-] *n.* 菊苣
celtuce['sɛljʊl] *n.* 青菜
kale[kel] *n.* 羽衣甘蓝；无头甘蓝
fennel[ˈfɛnəl] *n.* <植>茴香；小茴香
artichoke[ˈɑrtɪˌtʃok] *n.* 朝鲜蓟；洋蓟
bean[bin] *n.* 豆
pea[piː] *n.* 豌豆，豌豆类
red beet *n.* 甜菜根
wintermelon 冬瓜
bitter melon 苦瓜
curly endive 皱叶苦苣

Focus on Language

Dialogue 1

Steven(S): George, can you do me a favor?

George(G): Of course, Steven.

S: Help me to peel all of these tomatoes as soon as possible.

G: All of them?

S: Yes, all of them.

G: But how should I do?

S: Listen, firstly, cut an X into the bottom of the tomato, and be sure not too deep.

G: OK, and then?

S: After that, drop it into boiling water for about 10 to 15 seconds.

G: 10 to 15 seconds, okay.

S: Then, take it out of boiling water and plunge(投入) it into ice water. See, it is easy to use a paring knife to peel it now.

G: I see. I'll do my best.

S: Thank you very much.

G: You are most welcome.

Dialogue 2

Commis cook(C): What shall I do (with the ingredients of the dish)?

Vegetable chef(V): Wash the cucumbers, tomatoes, onions and lettuce well.

C: OK. The cucumbers are always dirty.

V: Soak it in salt water.

C: For how long?

V: For about 20 minutes.

C: And what shall I do then?

V: Cut the cucumbers into pieces, dice the tomatoes and onions.

C: How shall I cook the cucumbers?

V: For vegetable salad.

Commis cook: OK.

Task 1 Can you write down the English name below the pictures?

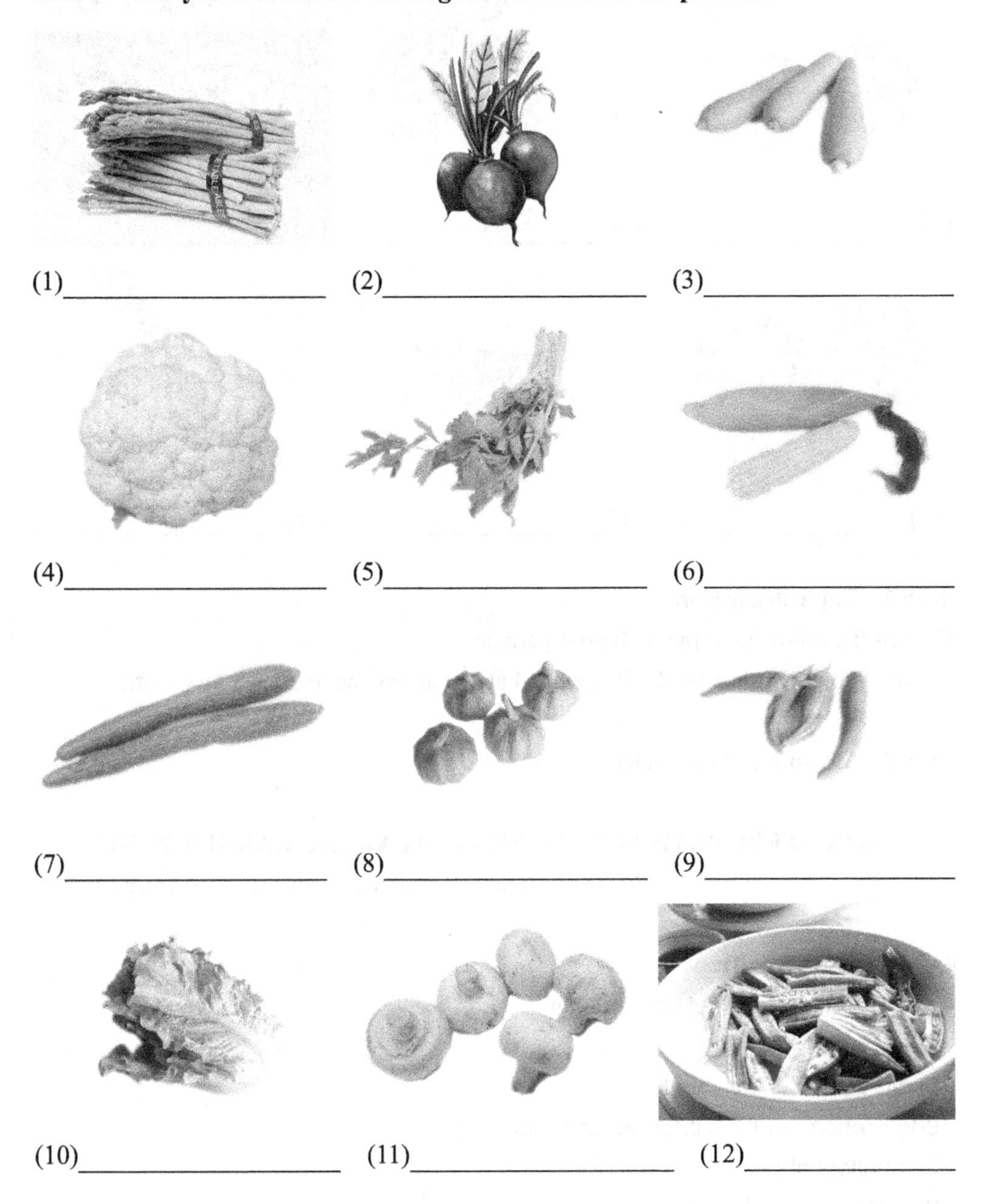

(13)______________ (14)______________ (15)______________

(16)______________ (17)______________ (18)______________

Task 2 Topic discussion.

Discuss the following topic with your partner.

Do you know what is a food critic, and what should you have to be a food critic?

Task 3 Try to retell the reciepe.

SMOKED EEL WITH FRENCH BEAN SALADAND CARROT PUREE

— A dish of beautiful textures: silky, fresh and soft

Ingredients(serves 4):

4×25g smoked eel fillet,without skin

For the French bean salad:

100g French beans,blanched and sliced

20g shallots,chopped

20ml white wine vinegar

2g Dijon mustard

20ml extra virgin olive oil

For the carrot puree:

100ml carrot juice

100g confit carrots(see basics)

20ml extra virgin olive oil

Direction:

Combine the mustard and vinegar in a mixing bowl. Emulsify with olive oil. Add the chopped shallots and beans then season to taste.

To make the carrot puree, pour the carrot juice into a saucepan and reduce by half.

Blend the carrot juice and confit carrot.Pass through a sieve and finish with olive oil.

Season to taste.

Garnish with horseradish cream, caviar and a sprig of chervil.

Task 4 Translate the following sentences into English.

1. 这份色拉是由生菜、西蓝花、土豆和芹菜做成的。

____

2. 洗花菜之前最好用盐水浸泡一下。

____

3. 请给我花椰菜而不要马铃薯泥可以吗?

____

4. 将 3 杯水煮沸后加盐及糖，将芦笋汆熟。

____

5. 你尝得出炖肉里有大蒜味儿吗？

____

6. 茄子是一种紫色的蔬菜。

____

7. 他们用沉重的木制压榨机把橄榄压碎。

____

Reading

American cooking style

Do you think American cooking is only about opening a bag of food and putting it into the microwave oven? If you think so, you are wrong. It is true that many Americans eat bread for breakfast, sandwiches for lunch and fast dinners. Many Americans like them because they can eat them in 10 minutes or less. But many Americans think that cooking skills are necessary. Parents — especially mothers — think it is important to let their children — especially daughters — learn how to cook. Most Americans think there's nothing better than a good meal cooked at home.

Every cook has his or her own cooking skills. But there are some skills that most people use. For example, Americans often bake to make food. In a big meal, there will be meat, a few vegetables, some bread and often a dessert. Americans also like to make the meal colorful. Having some different colors of food usually makes for a healthy meal.

Unit 5

Fruits

Lead in

Many hundreds of fruits, including fleshy fruits (like apple, kiwifruit, mango,peach, pear, and watermelon) are commercially valuable as human food, eaten both fresh and as jams, marmalade and other preserves. Fruits are also used in manufactured foods (e.g., cakes, cookies, ice cream, muffins, or yogurt) or beverages, such as fruit juices (e.g., apple juice, grape juice, or orange juice) or alcoholic beverages (e.g., brandy, fruit beer, or wine). Fruits are also used for gift giving, e.g., in the form of Fruit Baskets and Fruit Bouquets. Fresh fruits are generally high in fiber, vitamin C, and water. Regular consumption of fruit is generally associated with reduced risks of several diseases and functional declines associated with aging.

All fruits benefit from proper post harvest care, and in many fruits, the plant hormone ethylene causes ripening. Therefore, maintaining most fruits in an efficient cold chain is optimal for post harvest storage, with the aim of extending and ensuring shelf life.

Vocabulary

blueberry [ˈbluˌbɛri] *n.* 越橘

cherry [ˈtʃɛri] *n.* 樱桃

grape [grep] *n.* 葡萄

strawberry [ˈstrɔˌbɛri] *n.* 草莓

plum [plʌm] *n.* 李子
peach [pitʃ] *n.* 桃子
date [det] *n.* 枣
apricot [ˈeprɪˌkɑt] *n.* 杏仁
apple ['æpəl] *n.* 苹果
pear [pɛr] *n.* 梨
grapefruit [ˈɡrepˌfrut] *n.* 葡萄柚
lemon [ˈlɛmən] *n.* 柠檬
tangerine [ˌtændʒəˈrin, ˈtændʒəˌrin] *n.* 柑橘
lime [laɪm] *n.* 酸橙
banana [bəˈnænə] *n.* 香蕉
durian [ˈdʊriən, -ˌɑn, ˈdjʊr-] *n.* 榴莲果
longan [ˈlɑŋɡən, ˈlɔŋ-] *n.* 龙眼
lychee [ˈlitʃi] *n.* 荔枝
papaya [pəˈpɑjə] *n.* 番木瓜果
kiwifruit['ki:wi:'fru:t] *n.* 奇异果
mango [ˈmæŋɡo] *n.* 芒果
pineapple [ˈpaɪnˌæpəl] *n.* 菠萝
melon [ˈmɛlən] *n.* 甜瓜
blackberry[ˈblækˌbɛri] *n.* 黑莓
pomelo[ˈpɑməˌlo] *n.* 柚子，文旦
orange[ˈɔrɪndʒ,ˈɑr-] *n.* 桔子，橙子
mandarin [ˈmændərɪn] *n.* 柑橘
pomegranate[ˈpɑmˌɡrænɪt, ˈpɑmɪ-, ˈpʌm-, ˈpʌmɪ-] *n.* 石榴；石榴树
fig[fɪɡ] *n.* 无花果；无花果树
watermelon[ˈwɔtɚˌmɛlən, ˈwɑtə-] *n.* 西瓜

Focus on Language

Dialogue 1

Jack(J): Do you have any plans for dinner tonight?

Linda(L): No, I was thinking of putting a frozen pizza in the oven or something. How about you?

J: I was thinking maybe we could make dinner together tonight. What do you think?

L: I'm absolutely useless at cooking!

J: I could teach you how to cook something healthy. Frozen pizzas are so bad for you!

L: I know they aren't good, but they are cheap, convenient, and fairly tasty.

J: I recently saw a recipe for spicy chicken curry in a magazine. Maybe we could try that.

L: Yeah, why not. Do you have all the ingredients?

J: I bought all the ingredients this morning, so let's start!

L: What do we do first?

J: First, you need to wash the vegetables and then chop them into little pieces.

L: OK. Should I heat the wok?

J: Yes. Once it gets hot, put a little oil in it, add the vegetables and stir fry them for a few minutes.

L: What about the chicken?

J: That needs to be cut into thin strips about 3 cm long and then it can be stir fried on its own until it's cooked through.

L: How about the rice?

J: I'll prepare that. Do you prefer white rice or brown rice?

L: White rice, please. None of that healthy brown stuff for me!

Dialogue 2

Waiter(W): What kind of vegetable do you want?

Guest(G): Any recommendation?

W: Yes. What about potatoes? Mashed, boiled or baked?

G: Mashed potatoes. Some asparagus please.

W: And, soup or salad?

G: Oh, I'll try the cream of cauliflower.

W: Good. Anything to drink while you wait?

G: An iced water, please.

Task 1　Can you write down the English name below the pictures?

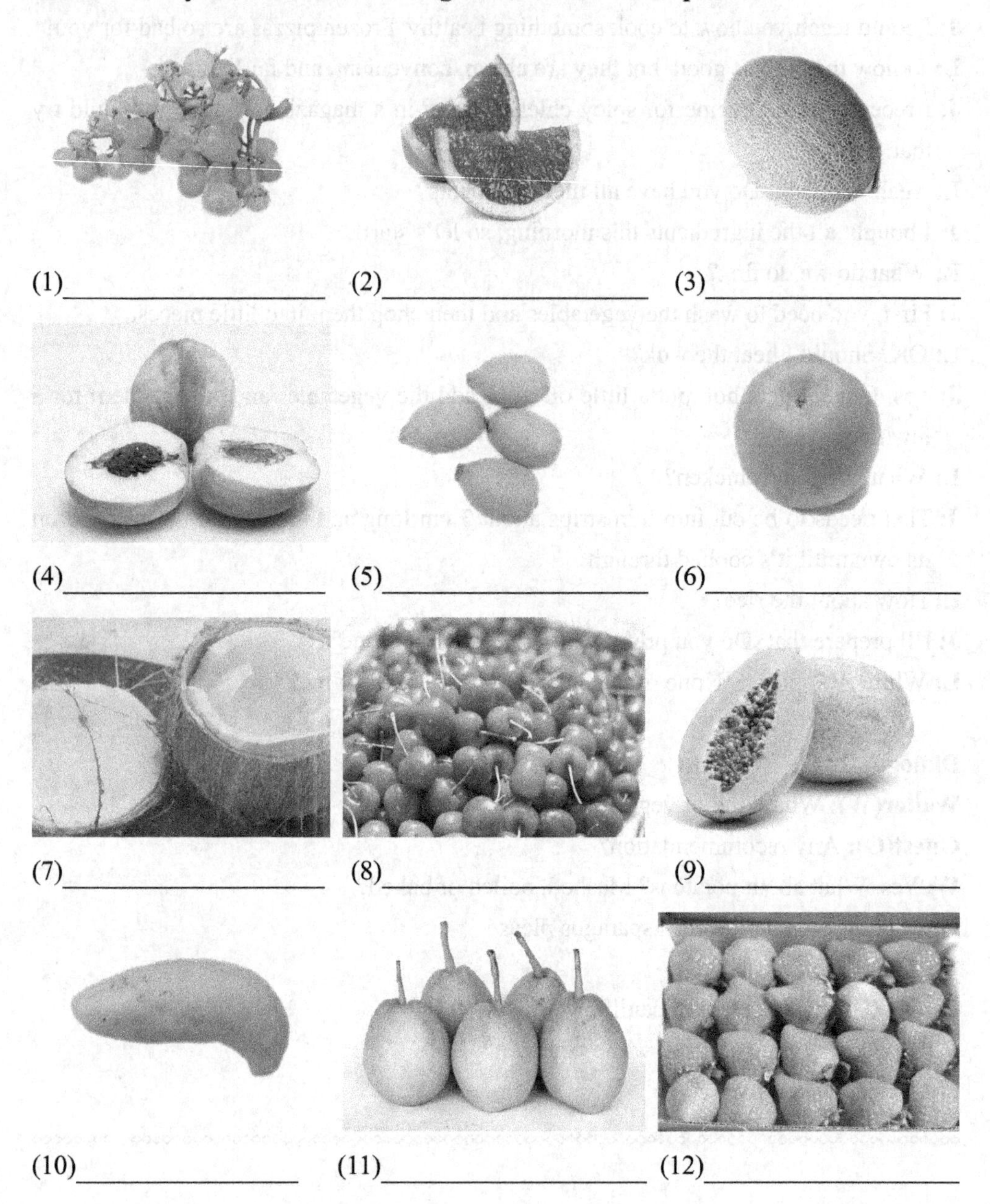
(1)________ (2)________ (3)________

(4)________ (5)________ (6)________

(7)________ (8)________ (9)________

(10)________ (11)________ (12)________

(13)____________________ (14)____________________ (15)____________________

Task 2 Try to retell the reciepe.

CONFIT DUCK LEG WITH ROCKET,APPLEAND BLUEBERRY SALAD

Tender and flavoursome, the dusk is complemented wonderfully by the peppery, fruity salad.

Ingredients(serves 4):

For the confit duck leg:

4 duck legs,trimmed
150ml duck fat
50ml duck jus
10g sea salt
2 sprigs of thyme
2 bay leaves
1 star anise
3g black peppercorns,crushed
20g dried apple skin
5g orange zest

For the rocket,apple and blueberry salad:

100g rocket
1 green apple
20g blueberries
20ml apple balsamic

Direction:

Marinate the duck legs in the sea salt,thyme, bay leaves, star anise, peppercorns, dried apple skin and orange zest for 24 hours. Remove the duck legs and rinse in cold water. Place the duck legs into a casserole dish and pour over the melted fat.Confit at 60 ℃ until tender. Remove from the fat and allow to cool. Before serving, reheat the duck legs inthe oven and glaze with the jus.

Prepare the salad by cutting the apple into strips. Combine with rocket and blueberries then dress with apple balsamic.Season to taste.

Task 3　Topic discussion.

(1) Which is your favourate fruit? Discribe the tasty of it.

(2) Discuss the common nutrition you could get from fruits.

Task 4　Translate the following sentences into English.

1. 浆果有营养并且有一种带甜味的口味。

____

2. 这种富含营养和奶油色的鳄梨被称为水果中的“巧克力”。

____

3. 杧果冰激凌的味道很不错。

____

4. 你所要做的就是把菠萝去皮然后挖出果心。

____

5. 切一颗奇异果，搭配火鸡肉、木瓜、杏仁片和菠菜叶，就是一道清爽的沙拉了。

____

6. 那盆鱼上配了几片柠檬作为装饰。

____

7. 厨房里好像正在做越橘果派。

____

8. 他把草莓酱涂在烤面包片上。

____

Reading

Apple crisp (name used in the United States and Canada) or apple crumble (name preferred in the United Kingdom, Australia and New Zealand) is a dessert consisting of baked chopped apples, topped with a crisp streusel crust.

Ingredients usually include cooked apples, butter, sugar, flour, cinnamon, and often oats and brown sugar, ginger, and/or nutmeg. One of the most common variants is apple rhubarb crisp, in which the rhubarb provides a tart contrast to the apples.

Many other kinds of fruit crisps are made. These may substitute other fruits, such as peaches, berries, or pears, for the apples.

Black bun is a type of fruit cake completely covered with pastry. It is Scottish in origin, originally eaten on Twelfth Night but now enjoyed at Hogmanay. The cake mixture typically contains raisins, currants, almonds, citrus peel, allspice, ginger, cinnamon and black pepper. It had originally been introduced following the return of Mary, Queen of Scots from France, but its original use at Twelfth Night ended with the Scottish Reformation. It was subsequently used for first-footing over Hogmanay.

A Brown Betty is a traditional American dessert made from fruit (usually apple, but also berries or pears) and sweetened crumbs. Similar to a cobbler or apple crisp, the fruit is baked and in this case the sweetened crumbs are in layers between the fruit. It is usually served with lemonsauce or whipped cream.

The dish was first mentioned in print in 1864. A recipe from 1877 uses apple sauce and cracker crumbs.

Apple Brown Betty was one of the favorite desserts of Ronald and Nancy Reagan in the White House.

An apple pie is a fruit pie in which the principal filling ingredient is apple. It is, on occasion, served with whipped cream or ice cream on top, or alongside cheddar cheese. The pastry is generally used top-and-bottom, making it a double-crust pie; the upper crust may be a circular or a pastry lattice woven of crosswise strips. Exceptions are deep-dish apple pie, with a top crust only, and open-face Tarte Tatin.

Unit 6

Meat and Poultry

Lead in

Meat can be broadly classified as "red" or "white" depending on the concentration of myoglobin in muscle fiber. When myoglobin is exposed to oxygen, reddish oxymyoglobin develops, making myoglobin-rich meat appear red. The redness of meat depends on species, animal age, and fibre type. Red meat contains more narrow muscle fibres that tend to operate over long periods without rest, while white meat contains more broad fibres that tend to work in short fast bursts.

Generally, the meat of adult mammals such as cows, sheep, goats, and horses is considered red, while chicken and turkey breast meat is considered white.

Vocabulary

beef [bif] *n.* 牛肉
flank [flæŋk] *n.* 侧面
brisket [ˈbrɪskɪt] *n.* 胸部
shank [ʃæŋk] *n.* 胫
pork [pɔrk, pork] *n.* 猪肉
bacon [ˈbekən] *n.* 培根
ham [hæm] *n.* 火腿
sausage [ˈsɔsɪdʒ] *n.* 香肠
turkey [ˈtə:ki] *n.* 火鸡
chicken [ˈtʃɪkən] *n.* 鸡
duck [dʌk] *n.* 鸭子
hen [hɛn] *n.* 母鸡

lamb [læm] n. 羔羊
veal [vil] n. 小牛肉
pancetta [pæn'setə] n. 未经熏制的咸肉
prosciutto [proˈʃuto] n. 意大利熏火腿
chorizo [tʃəˈrizo, -so] n. 西班牙加调料的口利左香肠
mortadella [ˌmɔrtəˈdɛlə] n. 熏香肠
cassoulet ['kæsʊleɪ] n. 豆焖肉
brochette [broˈʃɛt] n. 烤肉叉，小串烤肉
roast chicken 烤鸡
stuffed turkey 火鸡
deviled poussin 烤童子鸡
beef wellington 威灵顿牛肉馅饼
beef stew 红烩牛肉
lamb rib chop 羊肋排
ground meat 肉馅
blood sausage 血肠
skirt steak 肋排
rump roast 烤蹄髈，烤大腿部分
eye round roast 烤眼肉
top round roast 烤里仔盖
T-bone steak T 骨牛排
boneless top loin steak 腰脊牛排
tenderloin roast 烤菲力
porterhouse steak 上等腰肉牛排
back ribs 肋骨小排
top sirloin steak 沙朗牛排，菲力牛排
chuck eye roast 前肩肉眼 / 牛排
blade roast 煎颈片肉
smoked ham 熏火腿
Frankfurt sausages 法兰克福香肠
Genoa salami 热那亚式萨拉米香肠（无烟熏制的猪肉香肠）
goose foie gras 鹅肝酱
rump roast 烤蹄髈
Chinese dried sausage 中式香肠

Focus on Language

Dialogue 1

A: What's this?

B: It's a colander.

A: How can we use it?

B: It can be used as a strainer for draining vegetables and fruits.

A: Can it be used for washing as well?

B: Definitely.

A: What shall we do then?

B: We defrost the fish with the microwave first.

A: I see. What shall we use, pan or wok for fried fish?

B: Pans are not as handy as woks for Chinese cooking. Could you do me a favor to make sorbet with the blender?

A: My pleasure.

Dialogue 2

Jack(J): Good morning, Andy.

Andy(A): Good morning, Jack.

J: You look great!

A: Thank you.

J: What a nice day today! Let's start work. What kinds of material we get today?

A: We have beef steak with great quality from Australia.

J: OK. It must be really tender and juicy. It is suitable for grilling. Let's just marinate it for later use.

A: All right! I will preheat the oven.

Task 1 Try to write down the English name below the pictures.

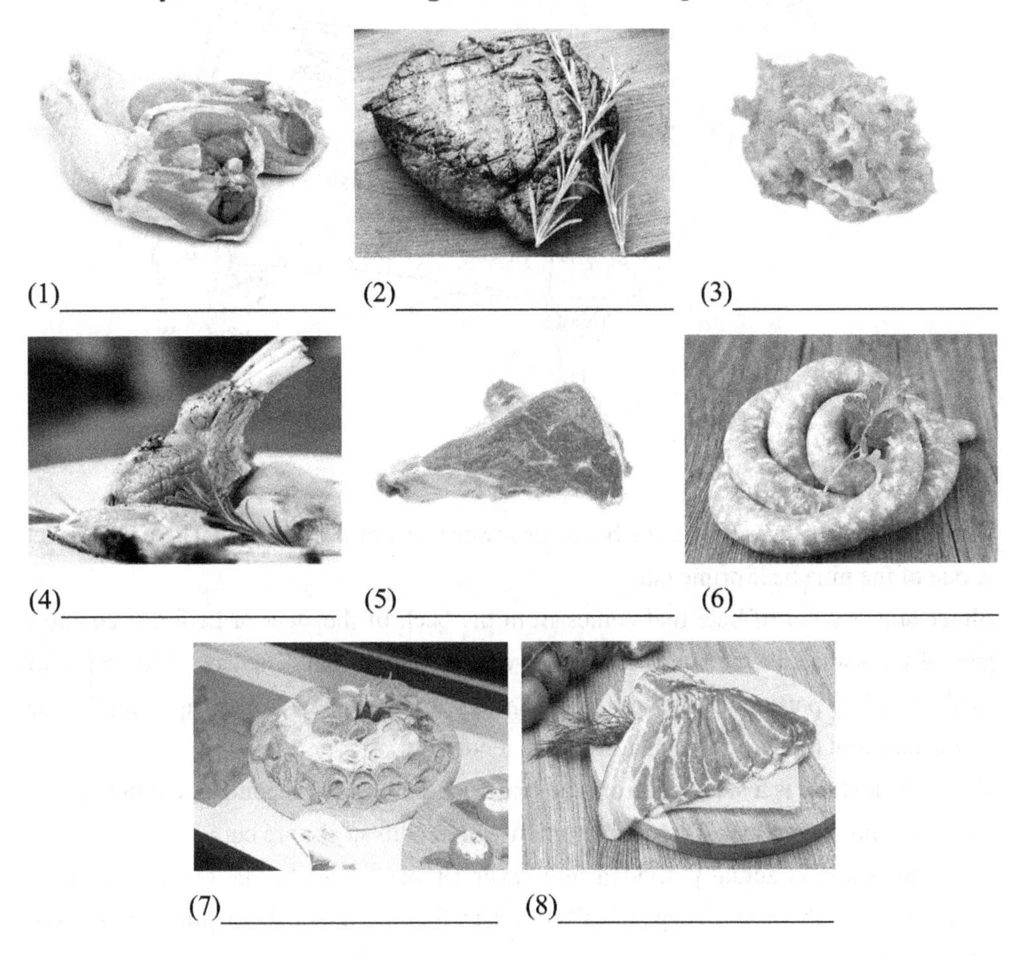

(1)____________ (2)____________ (3)____________

(4)____________ (5)____________ (6)____________

(7)____________ (8)____________

Task 2 Read the following passage based on the picture; discuss with your partner about the usage of every part of beef in cooking.

Chuck steak is a cut of beef and is part of the sub primal cut known as the chuck. The typical chuck steak is a rectangular cut, about 1 thick and containing parts of the shoulder bones, and is often known as a "7-bone steak". This cut is usually grilled or broiled.

Short rib sare a popular cut of beef. Beef short ribs are larger and usually more tender and meatier than their pork counterpart, pork spare ribs. Short ribs are cut from the rib and plate primal and a small corner of the square-cut chuck.

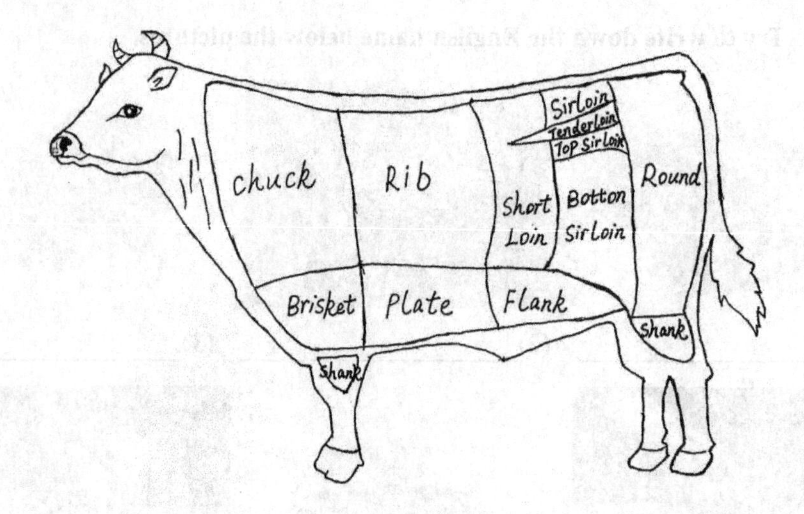

Brisket is a cut of meat from the breast or lower chest of beef or veal. The beef brisket is one of the nine beef prime cuts.

Short loin is a cut of beef that comes from the back of the steer or heifer. It contains part of the spine and includes the top loin and the tenderloin. This cut yields types of steak including porterhouse, strip steak (Kansas City Strip, New York Strip), and T-bone (a cut also containing partial meat from the tenderloin).

The sirloin steak is a steak cut from the rear back portion of the animal, continuing off the short loin from which T-bone, porterhouse, and club steaks are cut.

The sirloin is actually divided into several types of steak. The top sirloin is the most prized of these and is specifically marked for sale under that name. The bottom sirloin, which is less tender and much larger, is typically marked for sale simply as "sirloin steak." The bottom sirloin in turn connects to the sirloin tip roast.

A round steak is a steak from the round primal cut of beef. Specifically, a round steak is the eye (of) round, bottom round, and top round still connected, with or without the "round" bone (femur), and may include the knuckle (sirloin tip), depending on how the round is separated from the loin. This is a lean cut and it is moderately tough.

The flank steak is a beefsteak cut from the abdominal muscles of the cow. A relatively long and flat cut, flank steak is used in a variety of dishes including London broil and as an alternative to the traditional skirt steak in fajitas. It can be grilled, pan-fried, broiled, or braised for increased tenderness.

The beef shank is the shank (or leg) portion of a steer or heifer. In Britain the corresponding cuts of beef are the shin (the foreshank), and the leg (the hindshank).

Task 3 According to the picture below,try to describe the different conditions of cooked beef steak.

Rare，Very Rare：极生，煎的（Grill）时间不超过3分钟。外表有烧烤过的痕迹，但是里面还是冷的。切开时还有血水渗出，但是肉质极嫩，口感多汁。

Rare：生，煎的时间不超过4分钟。外表有烤焦痕迹，里面肉质呈现红色，但入口有热度。切开时有血水渗出，但是肉质极嫩，口感多汁。

Medium Rare：中生，煎的时间6~8分钟。外表有烧烤过的痕迹，里面已经全面加热，可以感受到热度，但是肉质还是红色。切开时还有稍许血水渗出，肉质嫩，口感多汁。

Medium：稍熟，通常说的5~6分熟。煎的时间8~10分钟。外表烧烤呈深褐色，里面除了中间部分粉红色外，外围部分呈现烧烤过的浅褐色。切开时流出褐色肉汁。

Medium Well：中熟，7分熟。煎的时间10~12分钟。外表烧烤呈深褐色，但是里面核心部分呈现少许红色外，外围部分呈现烧烤过的褐色。切开时流出褐色肉汁。

Well Done：全熟，煎的时间12~15分钟。外表已有明显烤焦痕迹，整片肉有热度，里面肉色呈现深褐色。

Task 4 Topic discussion.

Discuss the following topic about fast food with your partner.

(1) What do you think about fast food, such as high-calorie and time-saving?

(2) In what case will you choose fast food?

(3) Any problems caused by fast food?

Task 5 Translate the following sentences into English.

1. 我要吃水芹菜汤和牛排。

____

2. 打三个蛋，再加盐。

____

3. 他把小牛肉切成肉片。

____

4. 他把一片无骨牛排放进小圆面包里。

5. 将煎好的牛胸肉移放到一个大烤盘里，较肥面朝上。

____

6. 这是从牛的腰部嫩肉切下来的无骨肉排。

____

Task 6 Try to retell the recipes.

ROAST CHICKEN TACOS

Ingredients:

1 chicken

1 onion

1 avocado

1 lime

cilantro

salt and pepper

several patty

Direction:

Wash the chicken, season with salt and pepper.

Preheat the oven to 450 degree.

Roast the chicken at 450 degree for 35 minutes.

Cut the lemon by half.

Slice avocado into cubes, scoop the cubes out.

Chop the white onion into tots.

Chop the cilantro fine.

Grill the patty until it becomes golden brown.

Chop the chicken into slices.

Place onion, cilantro,avocado,and chicken on the patty.

Squeeze lemon on the top.

BBQ WINGS

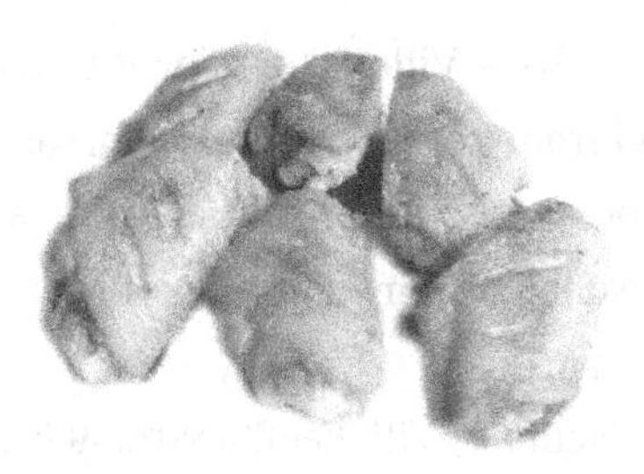

Ingredients:

10 chicken wings
1 teaspoon turmeric powder
1 teaspoon chili powder
1 tablespoon soy sauce
1 lemon grass -crushed
1 teaspoon sugar
1 tablespoon grounded peanuts
1 tablespoon oil
1 tablespoon cumin
1 red shallot-chopped

Direction:

Cut the chicken wings at the joints and marinate them with all the ingredients.
Leave for an hour.
Grill chicken wings in hot oven for about 1/2 hour.
Serve hot.

BEEF FILLET(SOUS VIDE COOKING)

Ingredients:

4 beef filet, about 2 inches (50mm) thick
56g unsalted butter or 1T (14g) per portion
2 shallots, each cut in half
4 sprigs of thyme

Kosher salt and coarse-ground black pepper, to taste olive oil or butter for searing

Directions:

For medium rare, set the Sous Vide Professional to the 135° (57.2℃), with rear pump flow switch closed and front flow switch set to full open.

Season beef filet portions with kosher salt and coarse ground black pepper. In a small vacuum bag, place seasoned, trimmed portion of beef filet with 1 T (14g) unsalted butter, half shallot and thyme sprig.

Seal portion to desired vacuum. With beef, a 90%~95% vacuum is desirable.

Once target temperature of 135 °F(57.2 ℃) is reached,place item in circulating water bath.

Cook to desired doneness, or about 60 minutes. You can hold at this temperature for up to 90 minutes without effecting quality or texture. Internal temperatures should reach a temperature of 135°F(57.2℃) for medium rare beef.

Remove the beef from vacuum bag. Dry off with paper towel. Season again, lightly, with kosher salt and coarse ground black pepper.

In a hot pan, grill or plancha, quickly sear off beef filet until browned. This adds additional flavor and texture and is known as Maillard reaction. The optimal result is to have even browning on all sides.

After 60 seconds of rest, beef may be sliced and plated.

Reading

Bacon

Bacon is a cured meat prepared from a pig. It is first cured using large quantities of salt, either in a brine or in a dry packing; the result is fresh bacon (also known as green bacon). Fresh bacon may then be further dried for weeks or months in cold air, or it may be boiled or smoked. Fresh and dried bacon is typically cooked before eating. Boiled bacon is ready to eat, as is some smoked bacon, but may be cooked further before eating.

Bacon is prepared from several different cuts of meat. It is usually made from side and back cuts of pork, except in the United States, where it is almost always prepared from pork belly (typically referred to as “streaky”, “fatty”, or “American style” outside of the US and Canada). The side cut has more meat and less fat than the belly. Bacon may be prepared from either of two distinct back cuts: fatback, which is almost pure fat, and pork loin, which is very lean. Bacon-cured pork loin is known as back bacon.

Bacon may be eaten smoked, boiled, fried, baked, or grilled, or used as a minor ingredient to flavor dishes. Bacon is also used for barding and larding roasts, especially game, e.g. venison, pheasant. The word is derived from the Old High German bacho, meaning "buttock", "ham" or "side of bacon", and cognate with the Old French bacon.

Ethics of eating meat

Ethical issues regarding the consumption of meat can include objections to the act of killing animals or to the agricultural practices used in meat production. Reasons for objecting to killing animals for consumption may include animal rights, environmental ethics, or an aversion to inflicting pain or harm on other sentient creatures. Some people, while not vegetarians, refuse to eat the flesh of certain animals, such as cows, pigs, cats, dogs, horses, or rabbits, due to cultural or religious traditions.

Some people eat only the flesh of animals which they believe have not been mistreated, and abstain from the meat of animals reared in factory farms or from particular products such as foie gras and veal. Some people also abstain from milk and its derivatives because the production of veal is a byproduct of the dairy industry. The ethical issues with factory farming relate to the high concentration of animals, animal waste, and the potential for dead animals in a small space. Critics argue that some techniques used in intensive agriculture can be cruel to animals. Foie gras is a food product made of the liver of ducks or geese that has been specially fattened by force feeding them corn. Veal is criticized because the veal calves may be highly restricted in movement; have unsuitable flooring; spend their entire lives indoors; experience prolonged sensory, social, and exploratory deprivation; and are more susceptible to high amounts of stress and disease.

Unit 7

Fish and Seafood

Lead in

Seafood is any form of sea life regarded as food by humans. Seafood prominently includes fish and shellfish. Shellfish includes various species of molluscs, crustaceans, and echinoderms. Historically, sea mammals such as whales and dolphins have been consumed as food, though that happens to a lesser extent these days. Edible sea plants, such as some seaweeds and microalgae, are widely eaten as seafood around the world, especially in Asia (see the category of sea vegetables). In North America, although not generally in the United Kingdom, the term "seafood" is extended to fresh water organisms eaten by humans. So all edible aquatic life may be referred to as seafood. For the sake of completeness, this article includes all edible aquatic life.

Vocabulary

eel [il] *n.* 鳗鱼
bass [beɪs] *n.* 鲈鱼
pike [paɪk] *n.* 长矛；梭鱼
carp [kɑrp] *n.* 鲤鱼
trout [traʊt] *n.* 鲑鳟鱼
mullet [ˈmʌlɪt] *n.* 胭脂鱼；鲻鱼
bluefish [ˈbluˌfɪʃ] *n.* 竹荚鱼类
shad [ʃæd] *n.* 西鲱，美洲西鲱
monkfish[ˈmʌŋkfiʃ] *n.* 安康鱼

sturgeon [ˈstɚdʒən] *n.* 鲟
caviar [ˈkæviˌɑː(r)] *n.* 鱼子酱
sardine [sɑrˈdin] *n.* 沙丁鱼
anchovy [ˈænˌtʃovi, ænˈtʃovi] *n.* 凤尾鱼；鳀鱼
cod [kɒd] *n.* 鳕鱼
salmon [ˈsæmən] *n.* 鲑鱼，大马哈鱼；鲑鱼肉
tuna [ˈtunə, ˈtju-] *n.* <鱼>金枪鱼（科），鲔鱼；金枪鱼罐头
shark [ʃɑrk] *n.* 鲨鱼
sole [sol] *n.* 鳎（可食用比目鱼）
scallop [ˈskɑləp, ˈskæl-] *n.* 扇贝；扇贝壳；扇（贝）形
mussel [ˈmʌsəl] *n.* 贻贝，蚌类；淡菜
oyster [ˈɔɪstɚ] *n.* 牡蛎
octopus [ˈɑktəpəs] *n.* 章鱼
cuttlefish [ˈkʌtlːˌfɪʃ] *n.* 乌贼，墨鱼
abalone [ˌæbəˈloni, ˈæbəˌlo-] *n.* <美>鲍鱼
shrimp [ʃrɪmp] *n.* 虾，小虾
crab [kræb] *n.* 蟹，蟹肉
lobster [ˈlɑbstɚ] *n.* 龙虾；龙虾肉
swordfish[ˈsɔrdˌfɪʃ] *n.* 旗鱼
haddock[ˈhædək] *n.* 小口鳕，黑线鳕（产于北大西洋的食用鱼）
turbot[ˈtəːbət] *n.* 大菱鲆
crayfish[ˈkreˌfɪʃ] *n.* 淡水螯虾（肉）；龙虾
gravlax [ˈgrɑvlɑks] *n.*（用盐、黑胡椒、小茴香、酒等腌制的）渍鲑鱼片
black pollock 黑鳕鱼
crab roulade 蟹肉卷
moules mariniere 烧贻贝
seared tuna 烤金枪鱼
sea bass *n.* <美>黑鲈，鲈科鱼的总称
pike-perch *n.* 狮鲈；梭鲈

Focus on Language

Dialogue 1

Waiter: Would you like to order now, sir?

Mark: Yes, I think so. Marian?

Marian: Yes, I'll have the salmon teriyaki, please.

Waiter: And what kind of potatoes would you like to go with that?

Marian: Baked, please. For the vegetable, I'd like broccoli.

Waiter: And would you care for soup or salad to start with?

Marian: I think I'll have a salad, please.

Waiter: All right. With what kind of dressing?

Marian: I'd like blue cheese.

Waiter: Yes. And you, sir? What will you have?

Mark: Those lobster tails on this menu sound pretty good.

Waiter: Oh, I'm very sorry, sir. We don't have any lobster now.

Mark: No lobster? OK…I guess I'll take the steak then. Rare.

Waiter: Yes. What about potatoes? Mashed, boiled or baked?

Mark: Mashed potatoes. For vegetable, I'd like asparagus.

Waiter: And, soup or salad?

Mark: Oh, I'll try the cream of cauliflower.

Waiter: Good. Anything to drink while you are waiting?

Marian: An iced water, please.

Mark: Make that two.

Dialogue 2

Commis: What should I do?

Chef: Cook the onion, please.

Commis: I already did.

Chef: Good. Then please mix the sesame seeds, water, garlic, salt, lemon juice, and red pepper.

Commis: How much lemon juice?

Chef: 20 tablespoons.

Commis: What is next?

Chef: Sprinkle the baking dish with breadcrumbs and parsley.

Commis: And then should I put the fish in the baking dish?

Chef: Yes. Pour the sesame seeds and onions over the fish.

Commis: Should I cover the fish?

Chef: No. Did you light the oven?

Commis: No.

Chef: Light the oven, please. Cook the fish at 400 degrees.

Commis: For how long?

Chef: For 20 to 25minutes.

Commis: After it is cooked, I will garnish the fish with the parsley and olives.

◇◇

Task 1 Try to write down the English name below the pictures.

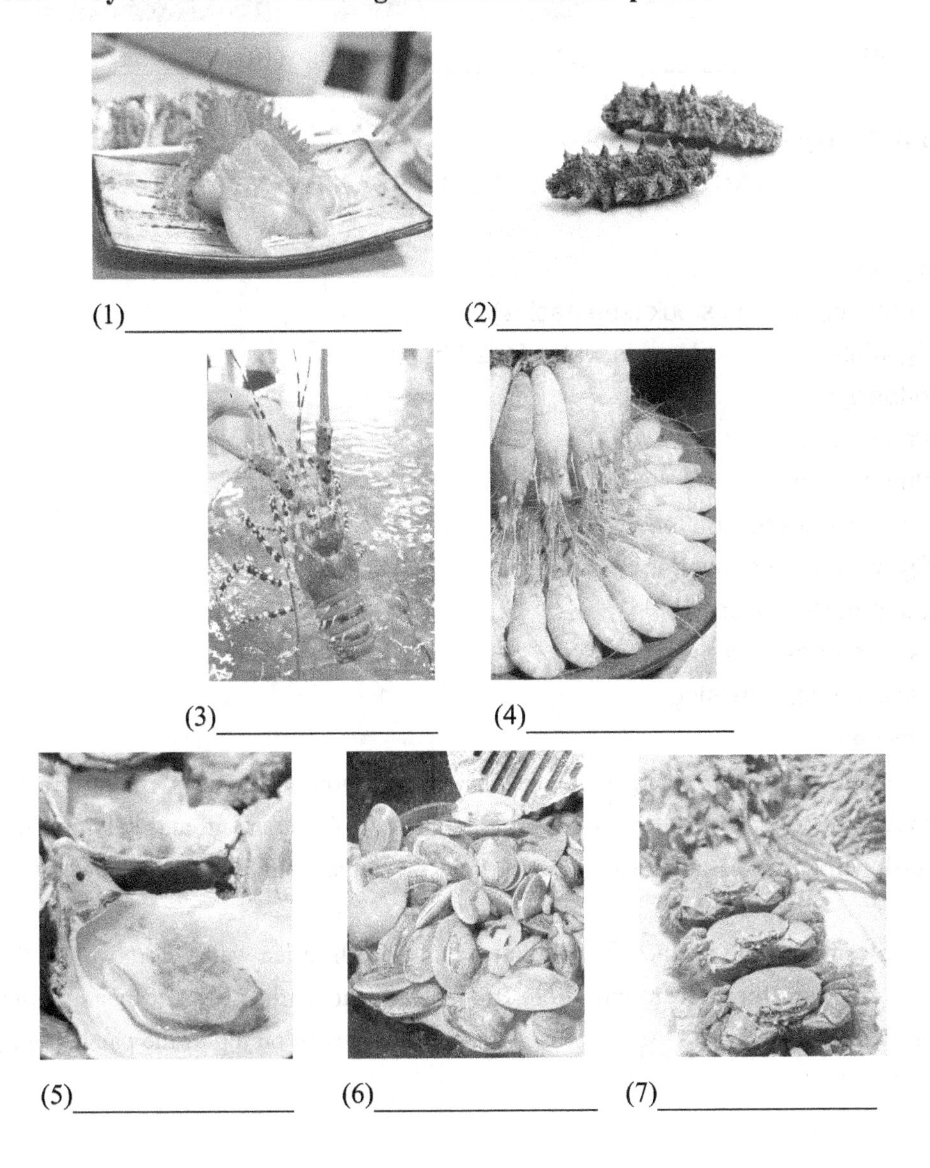

(8)________________ (9)________________ (10)________________

Task 2　Try to retell the recipes.

LANGOUSTINE AND WHITE BEAN PANNA COTTA

Ingredients:

200ml langoustine stock (see basics)
80g white beans, soaked in water for 24 hours
2g tarragon
20ml cream
20ml olive oil
1 gelatine sheet, soaked
30g white beans, cooked and peeled
20g tomato concasse
4 sprigs of baby cress
Prsces lemon dressing

Direction:

Cook the beans in langoustine stock with tarragon until tender. Blend in a food processor until smooth then pass through a sieve. Add the cream and gelatine sheet. Finish with olive oil and season to taste. Pour the panna cotta into glasses and leave to set.

Steam the langoustines for 3 minutes and refresh in ice water. Remove the flesh from the shell then clean and slice. Prepare a salad with the white beans,tomato concasse and baby cress. Adjust seasoning to taste. Present on top of the panna cotta to serve.

TUNA CARPACCIO WITH LOBSTER AND AVOCADO

Ingredients:

250g tuna, centre cut
2 avocados, peeled and cubed
150g lobster meat, cooked and cubed
5ml olive oil
10ml lemon juice
80g tomato concasse
20g shallots, chopped
2g basil, chopped
5ml lemon juice
2ml olive oil

Direction:

For the tuna Carpaccio, drizzle lemon oil on the tuna and season to taste. Wrap in cling film and freeze until required.

For the lobster and avocado salad, combine the ingredients in a mixing bowl and adjust seasoning. To prepare the tomato salsa, mix all the ingredients together. Again, season to taste.

To assemble, slice the tuna thinly and arrange on a plate. Divide the lobster and avocado salad equally into four round moulds. Top with the tomato salsa. Garnish with boiled quail eggs and caviar. Drizzle each plate with Pisces lemon aioli and balsamic reduction.

HONEY BAKED COD

Ingredients:

2 slice Cod
Marinade:
6 tablespoon Soy Sauce
5 tablespoon Honey
1 tablespoon Chinese Cooking Wine
1 tablespoon Water

Direction:

Mix the marinade well.
Pour over the the cod and marinate for 30 minutes.

Preheat oven at 160C.

Lay your baking tray with foil or Glad baking paper (which is what I use cos it doesn't stick to the fish).

Grill the cod for 20 minutes skin-side down before turning over.

Grill the other side for another 12 minutes or till fully cooked.

Serve immediately with salad.

MILK POACHED HALIBUT FLAVORED WITH ROSEMARY AND KAFFIR LIME LEAVES (Sous Vide cooking)

Ingredients:

4 portions of halibut — each 6 oz (170g) and 1-1/2 inch (40mm) thick

1/2 C(120ml) milk

3 sprigs rosemary

4 Kaffir Lime leaves, thinly sliced

Kosher salt, to taste

Direction:

Cooking time: 12-20 minutes

Serves: 4

Step one: Set the temperature on your Sous Vide Professional to 125 °F(51.7℃).

Step two: Season the halibut with salt. Place the halibut in a bag; add milk, rosemary and lime leaves and vacuum seal. Make sure that the fish is not overlapping in the bag.

Step three: Drop bag into the 125°F (51.7℃) water bath and make sure that it is completely submerged. If necessary, place a small weight on the bag to weight down the fish. After 11 minutes, remove the bag from the water and feel the fish for doneness. If the fish is not done, return the bag back to the water bath. Check every 2 minutes until the fish has reached desired doneness.

Step four: Gently remove halibut from vacuum bag. If a sear is desired, gently dry off the portion with paper or kitchen towel. Season as desired and sear in a hot pan with olive oil or butter. The halibut may also be grilled, if desired.

SALMON(Sous Vide cooking)

Ingredients:

4 center-cut salmon filets with medium fat content, pin bones removed, chilled -6oz (170g) or 3/2 inch (40mm) thick

1 T (15ml) extra virgin olive oil

1 bay leaf

Kosher salt and coarse ground black pepper, to taste Olive oil or clarified butter for searing

Direction:

Cooking time: 20 minutes

Serves: 4

Step one: Set the Sous Vide Professional to the desired temperature, with rear pump flow switch set to fully open. For medium rare salmon, 125°F (51.7℃) is found to be the best temperature if it has a medium level fat content.

Step two: In a small vacuum bag, place next to each other the seasoned, trimmed portions of salmon along with 1/2 T extra virgin olive oil and a half bay leaf.

Step three: Seal portion to desired vacuum. For delicate fish, the best vacuum percentage is 80%~90%. This will ensure the flesh of the fish portion is not compressed under vacuum, compromising the integrity of the delicate muscle fibers.

Step four: Once target temperature of 125°F (51.7℃) is reached, place salmon in circulating water bath.

Step five: Cook to desired doneness for 12~20 minutes. With salmon, the albumen, or white protein present in the fish, will begin to emerge from the flesh. Once this is barely visible, the fish is ready to remove from the bath.

Step six: Gently remove fish from vacuum bag. If a sear is desired, gently dry off the portion with paper or kitchen towel. Season as desired and sear in a hot pan with olive oil or butter.

Task 3 Translate the following sentences into English.

1. 鳕鱼已用盐腌起留着日后吃。

____

2. 黑线鳕通常烘焙，但有时涂大量黄油后烧烤。

____

3. 人们常常把大西洋鳕鱼或黑线鳕去骨切片后烹调。

____

4. 日本为鲔鱼主要消费国之一。

____

5. 先生，这是您点的清煎鲤鱼和烤牛肉。

____

6. 安康鱼配上橙子感觉非常精巧美妙。

____

7. 日本人喜欢吃生的鲑鱼肉。

____

8. 金枪鱼肉是可食用的肉，通常是装在罐头中或加工过的。

____

9. 扇贝洗净沥干水，每只开半，然后再切成小片。

____

Reading

Sous Vide Cooking

Chefs all over the world have enthusiastically embraced Sous Vide Cooking which relies on precise temperature control to achieve amazing flavor and texture. Food is vacuum-sealed and cooked at a gentle temperature in a precisely controlled water bath for perfect, repeatable results every time.

Although the term Sous Vide translates into "under vacuum", a better name for the technique would be precise temperature cooking. Food is packaged with a vacuum sealer in heat and food-grade plastic pouches before being cooked at a gentle temperature in a precisely controlled water bath. Removing air from the bag guarantees that the food is fully submerged in the water and evenly cooked from all sides.

Sous Vide Cooking is ideal for cooking delicate foods like fish and lobster. It's also great for retaining vibrant flavor and texture in vegetables or to enable lengthier cook tines on secondary cuts of meat without drying them out.

This new dimension in temperature control for your kitchen is easy to use and presents a completely different cooking experience.

Benefits of Sous Vide Cooking

(1) Culinary

Exact doneness for delicate foods;

Moist and tender texture;

Enhanced flavors;

Perfect results that are easy to repeat;

Retention of nutrients;

Ability to maintain food at serving temperature for an extended time without overcooking.

(2) Economic

Less shrinkage and up to 30% more yield;

Secondary cuts of meats turn out as tender as expensive primary cuts;

No waste from overcooking;

Ability to pre-cook and balance workload;

Perfect portion control;

Easy to learn and less training required.

Unit 8

Seasonings (herbs and spices) and Condiments

Lead in

Seasonings include herbs and spices, which are themselves frequently referred to as "seasonings". Seasoning includes a large or small amount of salt being added to a preparation. Salt may be used to draw out water, or to magnify a natural flavor of a food making it richer or more delicate, depending on the dish.

A condiment is a spice, sauce, or, preparation that is added to food to impart a particular flavor, to enhance its flavor, or in some cultures, to complement the dish. The term is originally described pickled or preserved foods, but it has shifted meaning over time.

Vocabulary

mayonnaise [ˌmeɪə'neɪz] *n.* 蛋黄酱；美乃兹

chutney['tʃʌtni] *n.* (印度)酸辣酱

dill [dɪl] *n.* 小茴香

anise [ˈænɪs] *n.* 茴芹

chervil [ˈtʃɚvəl] *n.* 山萝卜

rosemary [ˈrozˌmɛri] *n.* 迷迭香

oregano [əˈrɛɡəˌno, ɔˈrɛɡ-] *n.* 牛至

basil [ˈbæzəl,ˈbezəl] *n.* 罗勒属植物

sage [sedʒ] *n.* 鼠尾草

clove [klov] n. 丁香
nutmeg [ˈnʌtˌmɛg] n. 肉豆蔻
saffron [ˈsæfrən] n. 藏红花
curry [ˈkə:ri, ˈkʌri] n. 咖喱食品
turmeric [ˈtɚmərɪk] n. 姜黄
chili [ˈtʃɪli] n. 红辣椒
butter ['bʌtə] n. 黄油；奶油
mustard ['mʌstəd] n. 芥末；芥菜
dressing ['dresɪŋ] n. 调料
sauce [sɔ:s] n. 酱汁；调味汁
cumin seeds 孜然种子
vanilla pod 香草荚
white wine vinegar 白葡萄酒醋
rock sugar 冰糖
chili paste n. 辣椒酱
lemon balm [ˈlɛmən bɑm] 蜜蜂花
lemon grass [ˈlɛmən grɑs] 香茅草
lemon thyme [ˈlɛmən taɪm] 柠檬百里香
say sauce [sɔɪ sɔs] 酱油
soa salt [si sɔlt] 海盐
table salt [ˈtebəl sɔlt] 精盐
tomato ketchup 番茄酱

Focus on Language

Dialogue 1

Linda(L): Do you like cooking, Julia?

Julia(J): I really enjoy it, especially when it ends up tasting good!

L: How often do you usually cook?

J: I usually make a few salads for lunch throughout the week and I make dinner about 6 times a week.

L: That's a lot of cooking. You must save a lot of money by eating at home so much.

J: I do. If you cook at home, you can eat healthy food cheaply.

L: What kind of dishes do you usually make?

J: I always make either a beef roast or a chicken roast with asparagus, parsnips, peas, carrots and potatoes on Sunday.

L: Do you make a lot of traditional British food?

J: Aside from the Sunday roast, we usually eat bangers and mash, toad in a hole, or fish chips once a week.

L: How about spicy food?

J: My family loves spicy food. We often eat Chinese, Thai, Indian, or Mexican food when we're in the mood for spice.

L: What's your favorite dish to make?

J: I absolutely love making mousakka, which is a Greek dish with eggplant. But it takes a lot of time, so I don't often make it.

Dialogue 2

Commis (C): What shall I do with the fish?

Chef Brian (B): Scale it, cut off the fins and then gut it.

C: What kind of knife can I use to scale the fish?

B: Use the fish scaler to do it. And use these fish scissors to gut it.

C: So these are special scissors.

B: Yes. Use them to cut open the stomach of the fish. Then take out the guts.

C: OK.

B: After you finish it, cook some onion.

C: I already did.

B: Then mix the sesame seed, water, garlic, salt, lemon juice and red pepper with spoon.

C: All right. What next?

B: Sprinkle the baking dish with breadcrumbs and parsley.

C: Shall I put the fish in the baking dish?

B: Yes. Pour the sesame seed and onions over the fish. Light the oven and cook the fish at 400 degrees.

C: For how long?

B: For 20 or 25 minutes. When it is cooked, garnish the fish with parsley and olives.

Task 1 Try to write down the English name below the pictures.

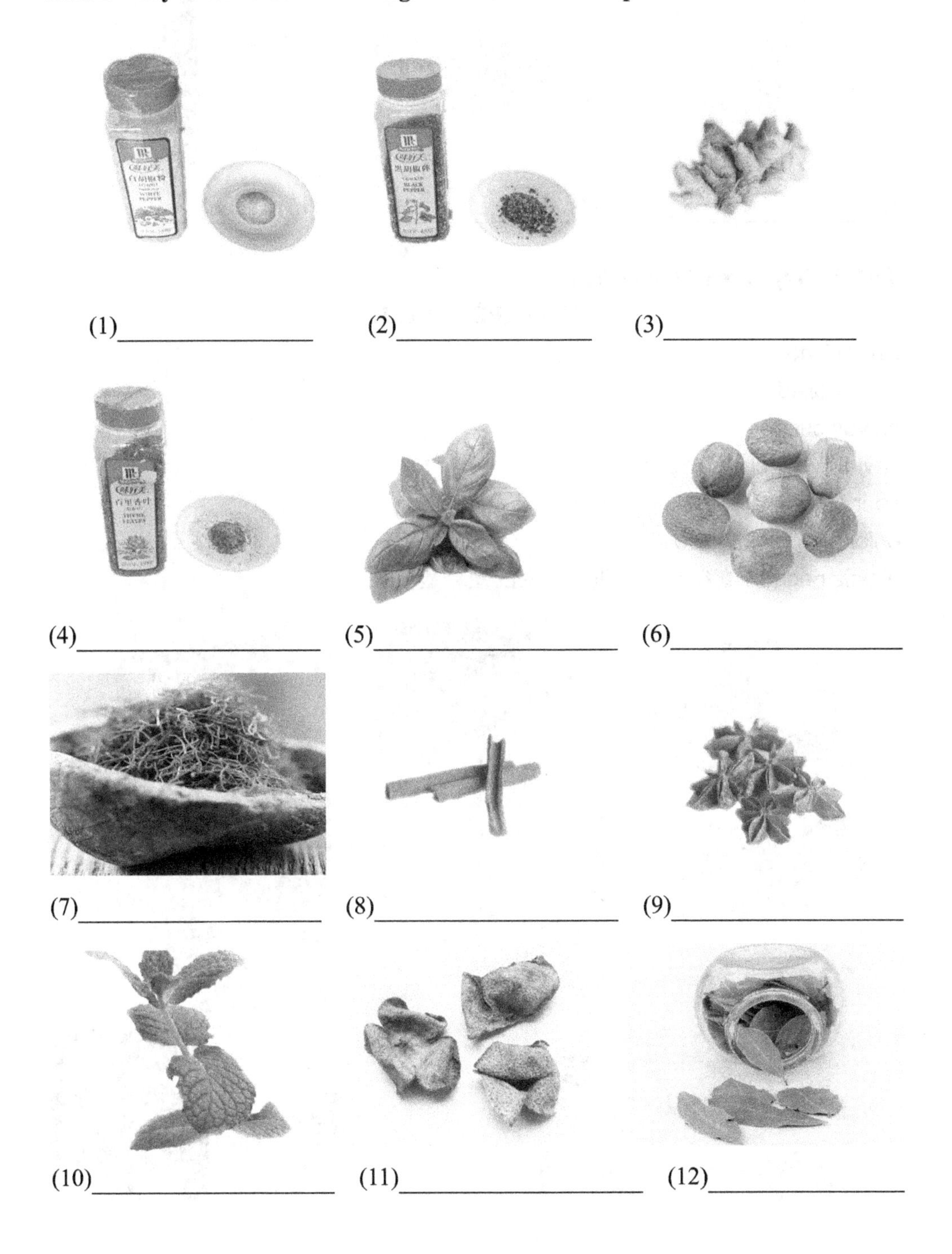

(1)________ (2)________ (3)________

(4)________ (5)________ (6)________

(7)________ (8)________ (9)________

(10)________ (11)________ (12)________

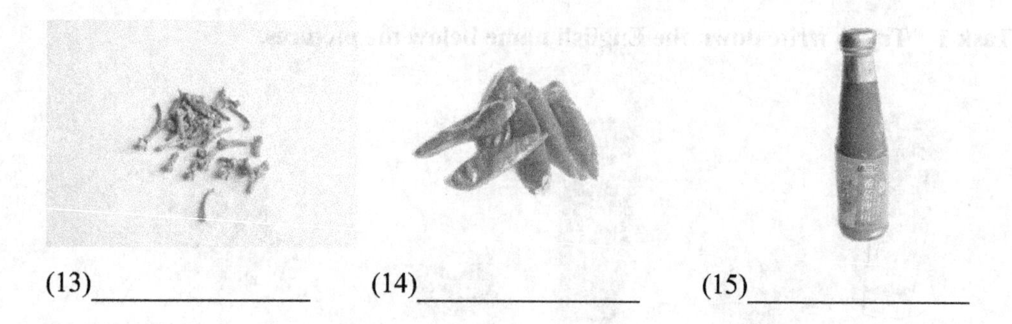

(13)________ (14)________ (15)________

Task 2 Try to retell the reciepe.

CANAPÉ CAVIAR

Ingredients:

eggs boiled

chives chopped

caviar (your choice)

sour crème

onions red finely diced

plastic ring (dia 直径 =3 cm, h 高 =5cm 塑料盛器)

SMOKED SALMON WITH DILL SOUR CREAM AND ASPARAGUS

Ingredients:

smoked salmon

French baguette

lettuce leaves

Thai asparagus

butter fresh

dill

sour cream

salt, pepper

Direction:

Cut baguette in a slight angle 1.5 cm thick, butter the slices evenly, put on some lettuce

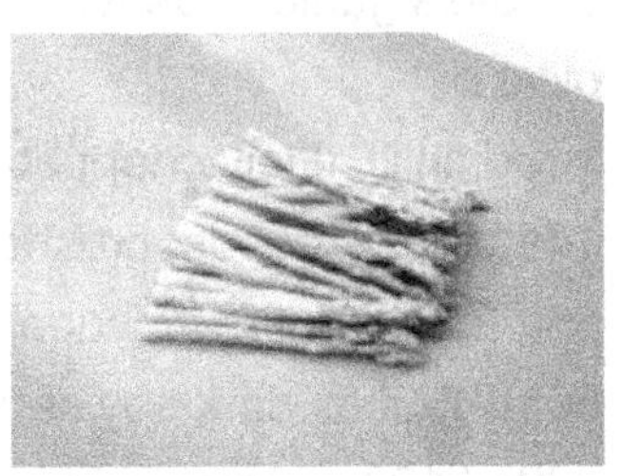

Clean off the dry skin of the salmon and slice salmon (if no pre sliced salmon available), put salmon slices on top of the lettuce, trim the asparagus and blanch quickly, refresh in ice water to maintain green color, trim to required length and put 3 pieces on top of the salmon

Season with pepper mill, make a sauce out of sour cream and chopped dill, season to taste and spoon on top of the asparagus, before serving

BBQ SAUCE

Ingredients:

1 small onion, chopped

3 cloves garlic, crushed

olive oil

1 red chilli, finely chopped

1 tsp fennel seeds, crushed

55g/2oz dark brown sugar

50ml/1¾fl oz dark soy sauce

300ml/10fl oz tomato ketchup

salt and pepper

Direction:

Fry the onion and garlic in olive oil with the chilli, fennel seeds and sugar.

Add the soy sauce and ketchup and season with salt and pepper.

Bring to the boil and simmer for a few minutes to combine the flavours. Use as a dip or to coat spare ribs, chicken or sausages.

WHITE SAUCE

Ingredients

25g/1oz butter

25g/1oz plain flour

600ml/1 pint milk

salt and white pepper

Direction:

Melt the butter in a saucepan.

Stir in the flour and cook for 1~2 minutes.

Take the pan off the heat and gradually stir in the milk to get a smooth sauce. Return to the heat and, stirring all the time, bring to the boil.

Simmer gently for 8~10 minutes and season with salt and white pepper.

Task 3 Translate the following sentences into English.

1. 鱼露是沿海地区的传统发酵调味品。

____

2. 然后你把作料放进面条里，再加点盐。

____

3. 日本芥末通常配生鱼片。

____

4. 这肉里应该用些盐和芥末调味。

____

5. 在法国，芥菜籽被浸透，然后再磨成浆糊状。

____

6. 酸辣酱可以混合在任何一种印度餐中，形成一种不同的口味。

____

7. 我想在上面放一些火腿、香肠、蘑菇、洋葱、橄榄和菠萝。

____

8. 盐是一种常用的食物防腐剂。

____

9. 肉豆蔻常用作食物中的香料。

____

10. 这个汤的特殊香味是因为藏红花粉。

____

Reading 1

Soy sauce

A condiment originally from china. Soy sauce occupies a preeminent position in the cuisines of Asian countries. Its Japanese name is shoyu. Traditionally, soy sauce,shoyu and tamari refer to the liquid formed during the manufacture of miso.

Traditional Chinese soy sauce is made using whole soybeans and ground wheat. It can be more or less dark depending on its age and whether caramel or molasses has been added.

Tamari is made exclusively using soybeans or soybean meal(the residue from pressing the beans when oil is extracted);therefore, it contains no cereal grain. It sometimes contains additives such as monosodium glutamate and caramel. Tamari is dark and has a thicker consistency. Shoyu is lighter in color than Chinese soy sauce and slightly sweet.

Soy sauce (Chinese or Japanese) contains some of the alcohol produced during the fermentation of the cereal grains, whereas tamari has none. The soy sauce found in supermarkets is usually a synthetic product that is a pale imitation of the original.

Reading 2

Culinary herbs

Culinary herbs are distinguished from vegetables in that, like spices, they are used in small amounts and provide flavor rather than substance to food.Many culinary herbs are perennials such as thyme or lavender, while others are biennials such as parsley or annuals like basil. Some perennial herbs are shrubs (such as Rosemary, Rosmarinus officinalis), or trees (such as Bay laurel, Laurus nobilis) – this contrasts with botanical herbs, which by definition cannot be woody plants. Some plants are used as both herbs and spices, such as dill weed and dill seed or coriander leaves and seeds. Also, there are some herbs such as those in the mint family that are used for both culinary and medicinal purposes.

Unit 9

Eggs and Cheese

Lead in

Cheese is a food derived from milk that is produced in a wide range of flavors, textures, and forms by coagulation of the milk protein casein. It comprises proteins and fat from milk, usually the milk of cows, buffalo, goats, or sheep. During production, the milk is usually acidified, and adding the enzyme rennet causes coagulation. The solids are separated and pressed into final form. Some cheeses have molds on the rind or throughout. Most cheeses melt at cooking temperature.

Hundreds of types of cheese from various countries are produced. Their styles, textures and flavors depend on the origin of the milk (including the animal's diet), whether they have been pasteurized, the butterfat content, the bacteria and mold, the processing, and aging. Herbs, spices, or wood smoke may be used as flavoring agents. The yellow to red color of many cheeses, such as Red Leicester, is produced by adding annatto. Other ingredients may be added to some cheeses, such as black pepper, garlic, chives or cranberries.

For a few cheeses, the milk is curdled by adding acids such as vinegar or lemon juice. Most cheeses are acidified to a lesser degree by bacteria, which turn milk sugars into lactic acid, then the addition of rennet completes the curdling. Vegetarian alternatives to rennet are available; most are produced by fermentation of the fungus Mucor miehei, but others have been extracted from various species of the Cynara thistle family. Cheesemakers near a dairy region may benefit from fresher, lower-priced milk, and lower shipping costs.

Vocabulary

cheese [tʃi:z] *n.* 干酪；乳酪；奶酪
cream [kri:m] *n.* 奶油；乳酪
frittata [fri'ta:tə] 菜肉馅煎蛋饼
butter ['bʌtər] *n.* 黄油；奶油
yogurt ['joʊgərt] *n.* 酸奶；酵母乳
pudding ['pʊdɪŋ] *n.* 布丁
omelet ['ɒmlɪt] *n.* 煎蛋；鸡蛋卷
blue cheese 蓝芝士
cream cheese 奶油奶酪
goat cheese 山羊乳干酪
whey cheese 乳清干酪
smoked cheese 熏制的奶酪
scrambled egg *n.* 炒蛋
egg benedict 烟肉及鸡蛋松饼
eggs Florentine 佛罗伦萨式蛋羹
quiche Lorraine 法式蛋塔
sour butter 酸黄油

Focus on Language

Dialogue 1

Commis: What kind of omelet should I fix?

Chef: A cheese omelet. Did you break three eggs?

Commis: No, not yet.

Chef: Break three eggs, please. Mix the eggs with salt and pepper.

Commis: I will heat the butter in the omelet pan.

Chef: Right.

Commis: The butter is browning.

Chef: Pour in the eggs…and stir quickly!

Commis: Should I use a wooden spoon?

Chef: No, use a fork.

Commis: What is next?

Chef: Sprinkle the omelet with cheese.

Commis: OK, now I will fold the omelet.

Chef: Right. Now slide the omelet into the plate.

Dialogue 2

W: What would you like for side dish?

G: Let me just think about it. Any recommendation?

W: How about mashed potatos?And what for vegetable?

G: I'd like asparagus.

W: And soup or salad?

G: Oh, I'll try the cream of cauliflower.

W: Good. Anything to drink while you are waiting?

G: Make two iced water,please.

Task 1 Look at the pictures below and try to translate the name of those dishes into Chinese.

cheesecake melted cheese swiss fondue

milk mousse

Task 2 Try to retell the recipes.

POACHED EGGS (SOUS VIDE COOKING)

Ingredients:

Large hen eggs, duck eggs or quail eggs—quantity is variable.

Direction:

Cooking time: 60 minutes

Step one: Set the rear pump flow switch to fully closed. Set the front flow switch to the minimum flow to ensure the delicate proteins in the whites do not separate from agitation.

Step two: Set the Sous Vide Professional to the desired temperature based on desired doneness of egg:143.5°F(62℃)for soft whites, 145.5°F (63 ℃) for medium set whites or 147.0°F (64 ℃) for firm set whites.

Step three: Once target temperature is reached, gently place eggs in circulating water bath. You may want to use a ladle or slotted spoon to gently lower the eggs so they do not crack.

Step four: Cook to desired doneness for 45 minutes. It's a general rule that most chicken and duck eggs will set to desired doneness in 60 minutes (approximately 1 minute/gm of egg). Quail eggs will generally cook to desired doneness in 20~30 minutes. The proteins will start to denature after 120 minutes, resulting in unpleasant textures.

Step five: If plating immediately, gently crack egg onto a paper towel to capture any excess liquid. Gently roll egg off of the towel onto plate.

If serving at a later point, plunge egg into ice bath. Store up to 48 hours under refrigeration. Reheat egg by placing in 140°F(60℃) circulating bath or placing cracked egg into a pot of simmering water for 60 seconds.

GOUGERES

Ingredients:

1 quantity cream puff pastry

1/3 cup finely shredded Gruyere or Cheddar cheese 1 egg, beaten

Direction:

Preparation time 25 minutes

Total cooking time 25 minutes

Makes 25~30

Preheat the oven to 32.5°F. Lightly grease two baking sheets. Mix half the cheese into the pastry dough.

Spoon the mixture into a pastry bag fitted with a small plain tip. Pipe out 1-inch balls of dough onto the prepared sheets, leaving a space of 1/4 inches between each ball. Using a fork dipped in the beaten egg, slightly flatten the top of each ball. Sprinkle with the remaining shredded cheese. Bake the balls for 20~25minutes, or until they have puffed up and are golden brown. Serve hot.

Chef's Tip:

This is a very simple and light finger food to serve with predinner drinks. Gougeres are sometimes served in restaurants with drinks and referred to as amuse-gueule, the French term for an appetizer.

GOAT CHEESE ESCABECHE WITH TRUFFLE GREEN PEA

A sharp, tangy bite that perefctly complements the smokiness of the char-grilled vegetables

Ingredients(serves 4):

For the goats' cheese escabeche:

1 carrot,peeled

1 green zucchini

1 yellow zucchini

1 eggplant

300g goats' cheese

Lemon juice

Olivia oil

Thyme

Salt and pepper

For the truffle green pea puree:

100g green peas,cooked

50ml vegetable stock

20ml truffle oil

For the truffle foam:

10ml truffle juice

80ml cooking cream

80ml skimmed milk

40ml vegetable stock

20g shallots,chopped

1 tablespoon truffle oil

Direction:

Slice the vegetables and marinate with salt, pepper, lemon juice, olive oil and thyme. Grill until tender and set aside to cool. Arrange the vegetables on a sushi mat.

Spread goats' cheese evenly on top and rollup. Cut into portions then warm in the oven when required.

Use a blender to puree the peas and vegetable stock then strain through a muslin cloth. Add the truffle oil and season to taste.

To prepare the truffle foam, sweat off the shallots. Add the vegetable stock and reduce to syrup. Pour in the milk and cream then reduce by half. Season to taste. Blitz with a blender to cream the foam.

To assemble place goats cheese escabeche onto a plate, top with a spoon of truffle foam and garnish with green pea puree.

Reading 1

Egg salad is part of a tradition of salads involving a high-protein and low-carbohydrate food mixed with seasonings in the form of spices, herbs, and other foods, and bound with mayonnaise. Its siblings include tuna salad, chicken salad, ham salad, lobster salad, and crab salad.

Egg salad is often used as a sandwich filling, typically made of chopped hard-boiled eggs, mayonnaise, mustard, minced celery, onion, salt, pepper and paprika. It is also often used as a topping on green salads.

A closely related sandwich filler is egg mayonnaise, where chopped hard-boiled egg is mixed with mayonnaise only.

Egg salad can be made creatively with any number of other cold foods added. Onions, lettuce, pickles, pickle relish, capers, bacon, pepper, cheese, celery and cucumber are common ingredients.

In Britain and Ireland, egg salad refers to green or mixed vegetable salad with egg on the side as the protein part of the meal, but not necessarily in mayonnaise.

Reading 2

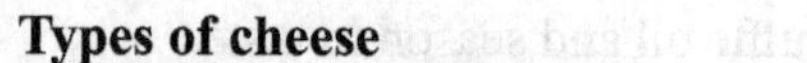

Types of cheese

Feta from Greece

There are many types of cheese, with around 500 different varieties recognised by the International Dairy Federation, over 400 identified by Walter and Hargrove, over 500 by Burkhalter, and over 1,000 by Sandine and Elliker. The varieties may be grouped or classified into types according to criteria such as length of ageing, texture, methods of making, fat content, animal milk, country or region of origin, etc. —with these criteria either being used singly or in combination, but with no single method being universally used. The method most commonly and traditionally used is based on moisture content, which is then further discriminated by fat content and curing or ripening methods. Some attempts have been made to rationalise the classification of cheese — a scheme was proposed by Pieter Walstra which uses the primary and secondary starter combined with moisture content, and Walter and Hargrove suggested classifying by production methods which produces 18 types, which are then further grouped by moisture content.

Moisture content (soft to hard)

Categorizing cheeses by firmness is a common but inexact practice. The lines between "soft", "semi-soft", "semi-hard", and "hard" are arbitrary, and many types of cheese are made in softer or firmer variations. The main factor that controls cheese hardness is moisture content, which depends largely on the pressure with which it is packed into molds, and on aging time.

Fresh, whey and stretched curd cheeses

The main factor in the categorization of these cheese is their age. Fresh cheeses without additional preservatives can spoil in a matter of days.

Content (double cream, goat, ewe and water buffalo).

Emmentaler

Some cheeses are categorized by the source of the milk used to produce them or by the added fat content of the milk from which they are produced. While most of the world's commercially available cheese is made from cows' milk, many parts of the world also produce cheese from goats and sheep. Double cream cheeses are soft cheeses of cows' milk enriched with cream so that their fat content is 60% or, in the case of triple creams, 75%.

Soft-ripened and blue-vein

There are at least three main categories of cheese in which the presence of mold is a significant feature: soft ripened cheeses, washed rind cheeses and blue cheeses.

Processed cheeses

Processed cheese is made from traditional cheese and emulsifying salts, often with the addition of milk, more salt, preservatives, and food coloring. It is inexpensive, consistent, and melts smoothly. It is sold packaged and either pre-sliced or unsliced, in a number of varieties. It is also available in aerosol cans in some countries.

Unit 10

Grains and Pastry

Lead in

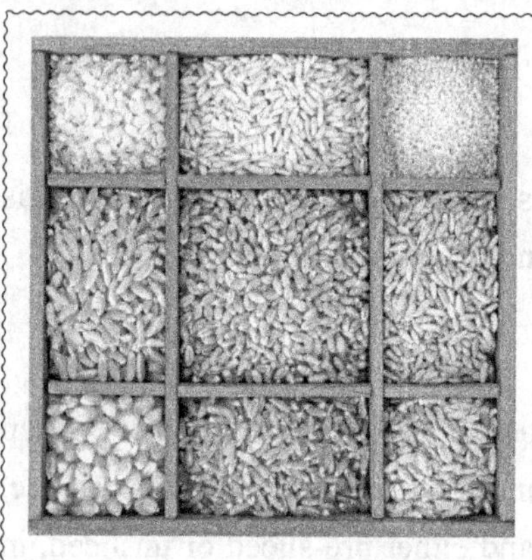

Grains are small, hard, dry seeds, with or without attached hulls or fruit layers, harvested for human or animal consumption. Agronomists also call the plants producing such seeds "grain crops". The two main types of commercial grain crops are cereals such as wheat and rye, and legumes such as beans and soybeans. Ubiquity of grain as a food source encouraged use of the term to describe other particles with volume or mass similar to an individual seed.

Pastry is a dough of flour, water and shortening that may be savoury or sweetened. Sweetened pastries are often described as *bakers' confectionery*. The word "pastries" suggests many kinds of baked products made from ingredients such as flour, sugar, milk, butter, shortening, baking powder, and eggs. Small tarts and other sweet baked products are called pastries. Common pastry dishes include pies, tarts, quiches and pasties.

Vocabulary

spaghetti [spə'geti] *n.* 意大利式细面条
pizza ['pi:tsə] *n.* 比萨饼
lasagna [lə'zɑ:njə] *n.* 烤宽面条
linguine [lɪŋ'gwi:ni:] *n.*(意大利)扁面条
ravioli [ˌrævi'oʊli] *n.* 馄饨
pasta ['pɑ:stə] *n.* 意大利面
cannelloni [ˌkænə'loʊni] *n.*(意大利的)烤碎肉卷子
paella [paɪ'elə] *n.*(西班牙)肉菜饭
risotto [rɪ'sɔ:toʊ] *n.* 意大利调味饭
wheat [wi:t] *n.* 小麦
flour ['flaʊər] *n.* 面粉；粉末
semolina [ˌsemə'li:nə] *n.* 粗粒小麦粉
couscous ['kʊskʊs] *n.* 蒸粗麦粉
millet ['mɪlɪt] *n.* 小米，黍
cornmeal ['kɔ:rnmi:l] *n.* 玉蜀黍粉；玉米片；麦片
tortellini [ˌtɔ:tə'li:nɪ] *n.*(意大利式)饺子
penne ['peneɪ] *n.* 短管状通心面
farfalle [fɑ:r'fæleɪ] *n.* 蝶形面食；蝴蝶面
rigatoni [ˌrɪgə'toʊnɪ] *n.* 波纹贝壳状通心粉
fusilli [fju:'sɪli] *n.* 螺旋形意大利面制品
fettucine ['fetəsɪn] *n.* 黄油酱汁面条；通心粉的一种
cannelloni [ˌkænə'loʊni] *n.*(意大利的)烤碎肉卷子
ziti ['zi:tɪ] *n.* 意大利通心面
macaroni [ˌmækə'roʊni] *n.* 通心粉
gnocchi ['njɑ:ki] *n.*(面粉或马铃薯做的)汤团；团子
spinach tagliatelle 菠菜意大利面
gnocchi 通心粉
rice cake *n.* 年糕
corn flour n. <美> 玉米粉； 玉米淀粉

Focus on Language

Dialogue 1

A: Have you ever tried Italian food?

B: Yes, several times already.

A: What did you eat?

B: Spaghetti.

A: How do you like it?

B: Well, you know I am used to eat Chinese food which is spicy.

A: Oh, you are exactly. I was puzzled to see some Chinese put extra spice onto Italian noodles. You know it can spoil the food.

B: But I understand them now.

Dialogue 2

Ms. Terry: Today's class is divided into two parts. First, I'll show you how to prepare pizza crust; then you will learn how to make pizza toppings. The ingredients for the dough are yeast, sugar, salt, vegetable oil, and flour.

Student 1: Is this self-rising flour?

Ms. Terry: Yes, it is. Now get a cup of warm water. Add yeast to the water and stir until it's dissolved. Then add one teaspoon of sugar, one teaspoon of salt, two teaspoons of oil, and two cups of flour.

Student 1: Should I keep stirring?

Ms. Terry: Yes. Add another half cup of flour and keep stirring. Now, sprinkle some flour on the board. Knead the dough until it feels elastic. While waiting for the dough to rise, let's prepare the toppings. Basically, you can put whatever you like on your crust. Here I have sliced Italian sausage, ground beef, peppers, onions, tomato paste, and shredded cheese.

Student 1: What kind of cheese is it?

Ms. Terry: We usually use Mozzarella cheese. Now set the dough on a greased pan, and layer your toppings onto the pizza dough. Put the whole thing in a preheated oven.

Student 2: How long should we bake it for?

Ms. Terry: For about 15 minutes.

Task 1 Try to write down the English names of the following pictures.

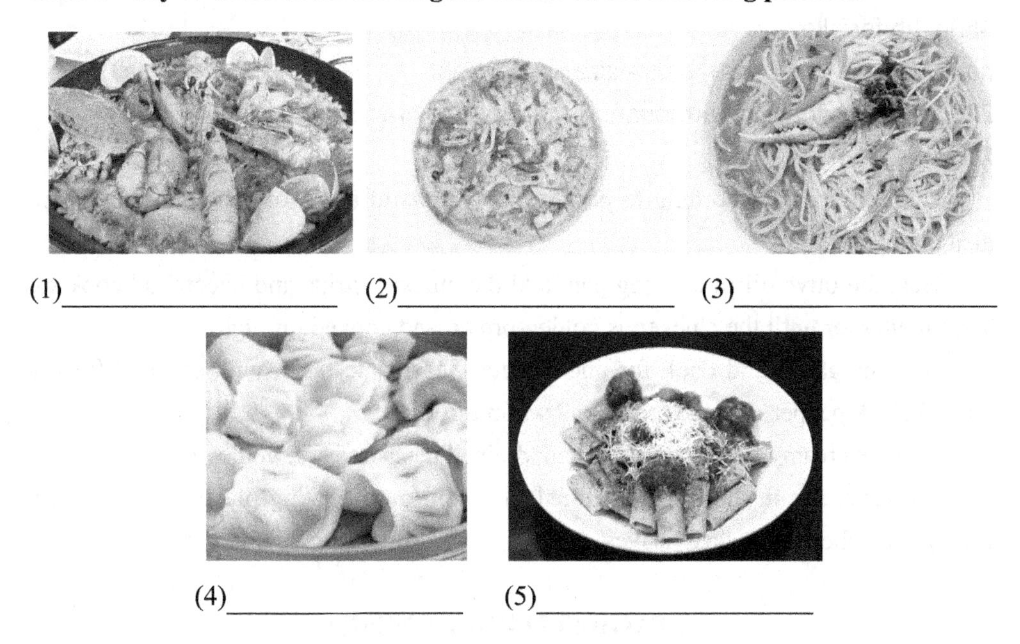

(1)________________ (2)________________ (3)________________

(4)________________ (5)________________

Task 2 Retell the recipes.

CREAMY CHICKEN, BACON AND BASIL PASTA

Ingredients:

3 tbsp olive oil

3 large boneless chicken thigh, skin removed, cut into strips

5 rashers bacon, chopped

1~2 garlic cloves, crushed

salt and freshly ground black pepper

300ml/10fl oz double cream
450g/1lb farfalle
handful fresh basil, torn, plus extra for garnish
200g/7oz cheddar or Parmesan, grated, plus extra to garnish

Direction:

Cook the pasta according to packet instructions in a pan of salted boiling water, then drain.

Heat the olive oil in a frying pan, add the chicken strips and bacon and cook for 3~4 minutes, or until the chicken is golden-brown and cooked through.

Add the garlic and cook for one minute. Season, to taste, with salt and freshly ground black pepper, add the cream and warm through.

Add the creamy sauce to the cooked, drained pasta and stir well.

To serve, stir in the basil and cheddar spoon onto serving plates. Garnish with extra grated cheese and basil leaves.

SPAGHETTI BOLOGNESE

Ingredients:

2 tbsp olive oil or sun-dried tomato oil from the jar
6 rashers of smoked streaky bacon, chopped
2 large onions, chopped
3 garlic cloves, crushed
1kg/2¼lb lean minced beef

2 large glasses of red wine
2×400g cans chopped tomatoes

1×290g jar antipasti marinated mushrooms, drained
2 fresh or dried bay leaves
1 tsp dried oregano or a small handful of fresh leaves, chopped
1 tsp dried thyme or a small handful of fresh leaves, chopped
Drizzle balsamic vinegar
12-14 sun-dried tomato halves, in oil
Salt and freshly ground black pepper
A good handful of fresh basil leaves, torn into small pieces
800g-1kg/1¾-2¼lb dried spaghetti
Lots of freshly grated parmesan, to serve

Direction:

Heat the oil in a large, heavy-based saucepan and fry the bacon until golden over a medium heat. Add the onions and garlic, frying until softened. Increase the heat and add the minced beef. Fry it until it has browned, breaking down any chunks of meat with a wooden spoon. Pour in the wine and boil until it has reduced in volume by about a third. Reduce the temperature and stir in the tomatoes, drained mushrooms, bay leaves, oregano, thyme and balsamic vinegar.

Either blitz the sun-dried tomatoes in a small blender with a little of the oil to loosen, or just finely chop before adding to the pan. Season well with salt and pepper. Cover with a lid and simmer the Bolognese sauce over a gentle heat for 1-1½ hours until it's rich and thickened, stirring occasionally. At the end of the cooking time, stir in the basil and add any extra seasoning if necessary.

Remove from the heat to "settle" while you cook the spaghetti in plenty of boiling salted water (for the time stated on the packet). Drain and divide between warmed plates. Scatter a little parmesan over the spaghetti before adding a good ladleful of the Bolognese sauce, finishing with a scattering of more cheese and a twist of black pepper.

PENNE WITH SPICY TOMATO AND MOZZARELLA SAUCE

Ingredients:

2 tsp olive oil
1 onion, peeled, finely chopped
3 garlic cloves, peeled, finely chopped
1 red chilli, seeds removed, finely chopped
1×400g/14oz can chopped tomatoes

splash red wine vinegar
1 tbsp tomato purée

Tabasco sauce, to taste
1 tsp sugar
small bunch fresh basil leaves, torn
salt and freshly ground black pepper
1×125g/4½oz ball fresh buffalo mozzarella, crumbled
400g/14oz dried penne pasta, cooked according to packet instructions, drained
grated parmesan, to serve

Direction:

Heat the oil in a saucepan over a medium heat. Add the onion and fry for 4~5 minutes, or until softened and beginning to colour.

Add the garlic and chilli and continue to fry for 1~2 minutes.

Pour in the chopped tomatoes, red wine vinegar, tomato purée, a splash or two of Tabasco (to taste), sugar and half of the basil leaves, and stir well. Season, to taste, with salt and freshly ground black pepper.

Bring the mixture to a simmer, then continue to simmer for 18-20 minutes, or until the sauce has thickened and reduced in volume.

Just before serving, stir the crumbled mozzarella into the spicy tomato sauce. Stir the drained pasta into the sauce.

To serve, divide the pasta and sauce equally among four serving plates. Sprinkle over the remaining torn basil leaves and the grated parmesan.

CHICKEN AND PEA RISOTTO

Ingredients:

For the risotto
olive oil, for frying
2 onions, sliced
400g/14oz arborio risotto rice
splash white wine
1 litre/1¾ pint hot vegetable stock
salt and freshly ground black pepper
200g/7oz fresh peas, blanched
200g/7oz asparagus, blanched
400g/14oz baby spinach
1 tbsp chopped fresh basil
75ml/2½fl oz double cream

For the chicken
oil, for frying
salt and freshly ground black pepper
4 chicken breasts, skin on
chicken or vegetable stock, to cover
parmesan shavings, to serve

Direction:

Preheat the oven to 200C/400F/Gas 6.

For the risotto, heat the oil in a pan and fry the onion until soft. Add the rice and stir well, then add the wine and simmer until reduced completely. Add a good ladleful of the hot stock and stir continuously. When all this has been absorbed, add more stock.

Continue adding more stock, stirring continuously, until the rice is cooked.

Season with salt and freshly ground black pepper and stir in the peas, asparagus, spinach, basil and cream.

For the chicken, season the chicken breasts well with salt and freshly ground black pepper. Heat the oil in a frying pan and place the chicken breasts skin-side down. Fry the chicken breasts on both sides until lightly browned.

Place the chicken breasts in an ovenproof dish and pour in enough stock to come one third of the way up the sides of the chicken breasts. Place in the oven for 15 minutes, or until completely cooked through.

To serve, place the risotto into serving bowls. Slice the chicken and place on top of the risotto, then sprinkle with parmesan shavings.

PAELLA

Ingredients:

170g/6oz chorizo, cut into thin slices
110g/4oz pancetta, cut into small dice
2 cloves garlic finely chopped
1 large Spanish onion, finely diced
1 red pepper, diced
1 tsp soft thyme leaves
¼ tsp dried red chilli flakes
570ml/1pint calasparra (Spanish short-grain) rice
1 tsp paprika

125ml/4fl oz dry white wine
1.2 litres/2 pints chicken stock, heated with ¼ tsp saffron strands
8 chicken thighs, each chopped in half and browned

18 small clams, cleaned
110g/4oz fresh or frozen peas
4 large tomatoes, de-seeded and diced
125ml/4fl oz good olive oil
1 head garlic, cloves separated and peeled
12 jumbo raw prawns, in shells
450g/1lb squid, cleaned and chopped into bite-sized pieces
5 tbsp chopped flatleaf parsley
Salt and freshly ground black pepper

Direction:

Heat half the olive oil in a paella dish or heavy-based saucepan. Add the chorizo and pancetta and fry until crisp. Add the garlic, onion and pepper and heat until softened. Add the thyme, chilli flakes and calasparra rice, and stir until all the grains of rice are nicely coated and glossy. Now add the paprika and dry white wine and when it is bubbling, pour in the hot chicken stock, add the chicken thighs and cook for 5~10 minutes.

Now place the clams into the dish with the join facing down so that the edges open outwards. Sprinkle in the peas and chopped tomatoes and continue to cook gently for another 10 minutes.

Meanwhile, heat the remaining oil with the garlic cloves in a separate pan and add the prawns. Fry quickly for a minute or two then add them to the paella. Now do the same with the squid and add them to the paella too.

Scatter the chopped parsley over the paella and serve immediately.

Unit 11

Kitchen Utensils

Lead in

A kitchen utensil is a hand-held, typically small tool or utensil that is used in the kitchen, for food-related functions. A cooking utensil is a utensil used in the kitchen for cooking. Other names for the same thing, or subsets thereof, derive from the word "ware", and describe kitchen utensils from a merchandising (and functional) point of view: kitchenware, wares for the kitchen; ovenware and bake ware, kitchen utensils that are for use inside ovens and for baking; cookware, merchandise used for cooking; and so forth.

Vocabulary

chopper ['tʃɒpə(r)] *n.* 斧头；砍刀
filter ['fɪltə(r)] *n.&v.* 过滤；渗透；过滤器
skimmer ['skɪmə] *n.* 撇取浮物的器具；网勺
colander ['kʌləndə(r)] *n.* 滤器；漏勺
chinois [ʃiː'nwɑː] *n.*（厨房用）漏勺
layer ['leɪə(r)] *n.&v.* 层；分层
spatula ['spætʃələ] *n.* 铲；抹刀；压舌片
mallet ['mælɪt] *n.* 木槌
escalope ['eskəlɒp] *n.* 薄肉片；（裹面包屑和鸡蛋的）炸肉块
stockpot ['stɒkpɒt] *n.* 锅子
saucepan ['sɔːspən] *n.* 长柄深锅

spider ['spaɪdə(r)]	*n.* 长柄平底锅
oven ['ʌvn]	*n.* 烤箱；烤炉
whisk [wɪsk]	*n.&v.* 打蛋器；搅拌
grater ['greɪtə(r)]	*n.* 擦子
grinder ['graɪndə(r)]	*n.* 磨工；研磨器
steel [sti:l]	*n.&v.&adj.* 钢；钢铁；使……硬如钢；钢的；坚强的
casserole ['kæsərəʊl]	*n.* 砂锅
toaster ['təʊstə(r)]	*n.* 烤面包机
microwave ['maɪkrəweɪv]	*n.&vt.* 微波；微波炉；用微波炉加热
steamer ['sti:mə(r)]	*n.* 蒸笼
deep fryer	油炸锅
knife set	套刀
oyster knife	牡蛎刀
boning knife	去骨刀
grapefruit knife	葡萄柚刀
carving knife	雕刻刀
cheese knife	起司刀
fish scissor	鱼剪刀
poultry shear	家禽剪
frying basket	油炸篮
conical strainer	圆锥形过滤器
palette knife	调色刀
basting brush	涂油刷
mixing bowl	搅拌碗
cutlet bat	把肉拍平的木板或木槌
saute pan	炒锅
stew pan	炖锅
roasting tray	烤盘
bone saw	骨锯
hot pad	隔热垫
hot glove	隔热手套
egg boiler	煮蛋器
dish washer	洗碗机
frying pan	煎锅；长柄平锅

List of key expressions

询问是否使用滤网　Did you clean the oil with a filter?

使用漏勺　You can use a skimmer, please.

放进油炸篮　Put the potato chips in a frying basket.

放进滤锅　Put them in a colander.

询问筛网用处　① What is a sieve for?

② Should I sift this flour with a sieve?

请人用圆锥形滤网　Use a conical strainer.

吩咐拿锥形漏勺　Give me a chinois, please.

吩咐用烤叉　Lift the meat with the roasting fork.

吩咐拿烤肉叉　Give me some skewers, please.

询问刀具　① Is my cook's knife for special jobs?

② What are these knives for?

③ What else can I use kitchen cutters for?

刀具用处　① Your cook's knife is for many different things.

② You can cut many different things with them.

③ Kitchen cutters are very useful.

④ This is a boning knife. A boning knife is used to remove bones from meat.

⑤ It's an oyster knife. It's for opening oyster shells. Oysters are a kind of shellfish.

询问可否用刀具去鱼鳞　Can I scale it with this knife?

用鱼剪取出内脏　Yes. And use these fish scissors to gut it.

抹糖霜　Let's put a layer of chocolate icing on the cake.

用抹刀　① With a palette knife.

② Use this palette knife.

搅拌材料　Stir the ingredients in the bowl.

询问用什么工具搅拌　With what should I stir the ingredients?

告知使用的各类汤匙　① Use a wooden spatula.

② Use a spoon.

③ Use a serving spoon.

询问可否使用漏眼匙　Can I serve peas with a slotted spoon?

请人拿长柄勺　① Give me a ladle, please.

② I need a big ladle. I don't have one.

吩咐用拍肉板打肉　Beat the meat with a culet bat.

询问是否为煎锅　Is this a frying pan?

告知为特殊的煎锅　① Yes. But it is a special type of frying pan.

② What is this frying pan used for?

③ It is a small frying pan for crepes.

使用热油　Sear the meat in hot fat.

使用烤箱　Cook it in the oven.

使用焖锅　Cook it in a braising pan.

告知小火慢炖鱼　Simmer the fish slowly.

告知热油　Heat some cooking oil in it.

询问是否要放进烤箱　Should I put the roasting tray in the oven?

使用打蛋器　① Use an eggbeater/ a whisk.

② Beat the ingredients with an eggbeater/ a whisk.

询问研磨器　Where is the grater?

将骨头对半切开　This is a big bone. I want to cut it in two.

Focus on Language

Dialogue 1

Michel(M): Wow! Your kitchen is awesome, Jane.

Jane(J): Thank you. It's newly renovated.

M: Oh, you've bought a new coffee maker.

J: Yes. Let's brew some coffee with it.

M: Great. How long shall we wait?

J: About 15 minutes. Michel, do you like red bean soup?

M: Yes, it's so cool.

J: We can cook it now.

M: It'll save time if we use the pressure cooker.

J: Oh, what a pity! My pressure cooker can't work now. But I can borrow one from my neighbor.

M: OK.

Dialogue 2

A: Chef, do we need to pan fry the lamb steak?

B: Sure, someone orderd a lamb steak meal just now,including potato mash and salad. Preheat the oven to 200 degree.

A: All right! How about the potato?

B: We need to make potato mash, peel it first.

A: OK, I need my paring knife.

B: Pass me frying pan and spatula, I will pan frying the steak.

A: OK. And the potato is done. I will mash it with masher and add some butter.

B: Well done.

Task 1 Look at the pictures and try to write down the English names below them.

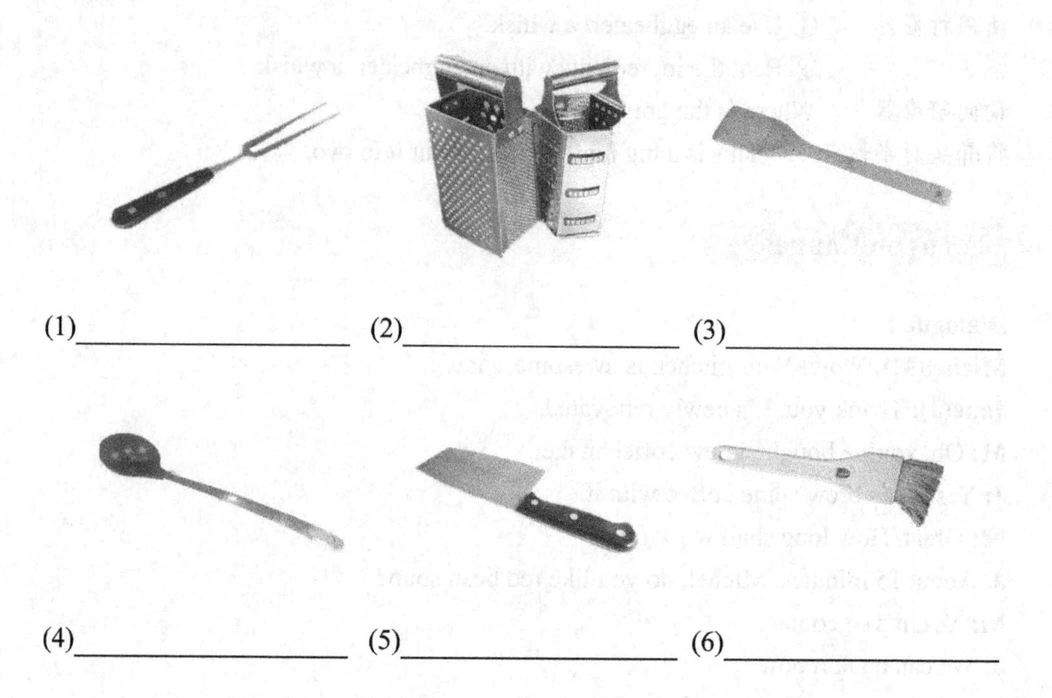

(1)________________ (2)________________ (3)________________

(4)________________ (5)________________ (6)________________

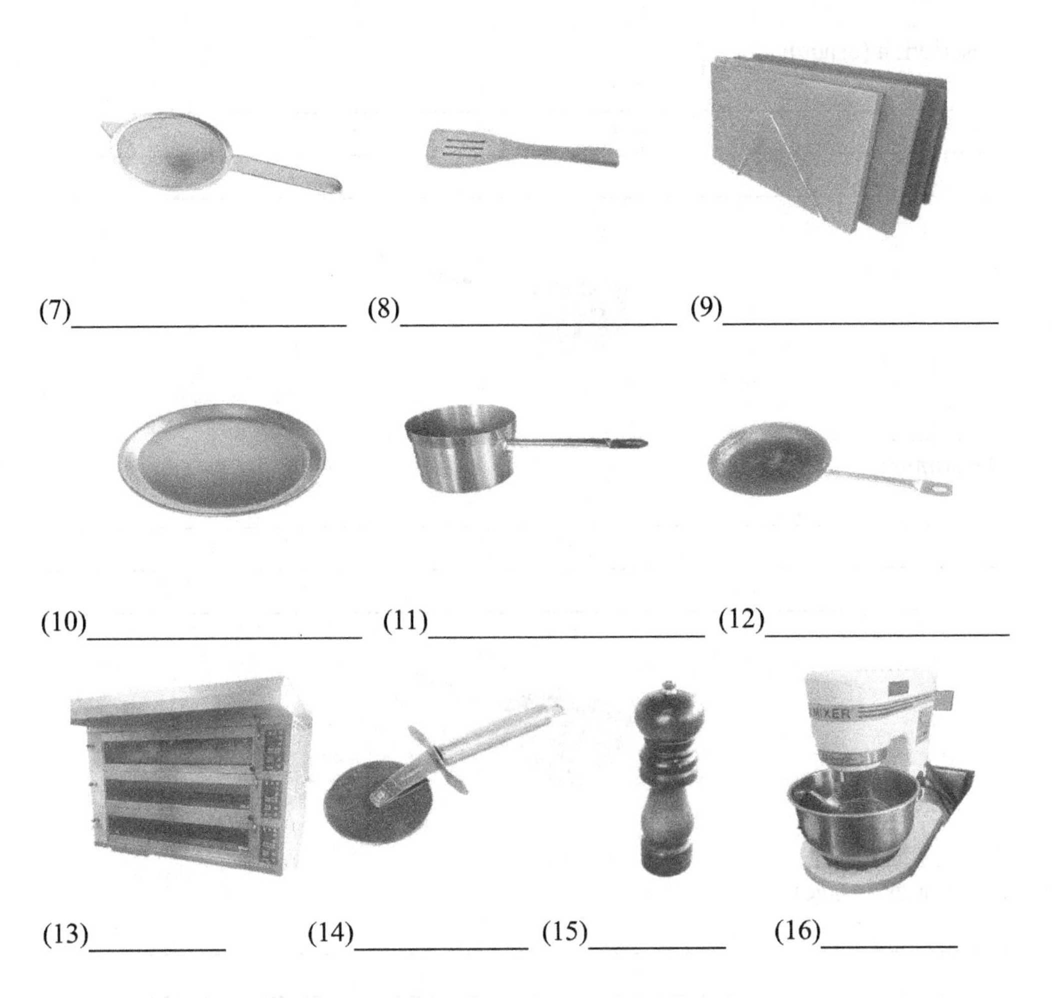

(7)________ (8)________ (9)________

(10)________ (11)________ (12)________

(13)________ (14)________ (15)________ (16)________

Task 2 Pair work. Look at the following pictures and figure out their names in English. Then try to tell what they are used for.

1. ________

Description (usage):

____

____

____

2. ______________

Description:

____

____

____

3. ______________

Description (usage):

____

____

____

4. ______________

Description (usage):

____

____

____

5. ________________

Description (usage):

____

____

____

Task 3 Different knives and scissors for preparing for different food (like meat, fish, vegetables, fruit, ect.). Match the following knives or scissors in column A with the food in column B.

Column A	Column B
a. pallet knife	1. chicken
b. grapefruit knife	2. beef
c. fish scissors	3. cake
d. cleaver	4. bone
e. cheese knife	5. orange
f. bone saw	6. fish
g. poultry shears	7. cheese
h. oyster knife	8. meat
i. steak knife	9. oyster

Task 4 Try to translate the following sentences into English.

1. 我要用筛网来筛这些面粉吗?

____

2. 请问你用圆锥形滤网来干什么?

____

3. 我要用漏勺把油表层清干净。

____

4. 我需要一个锥形漏勺，能麻烦你递给我吗?

____

5. 请问这把奇特的小刀用途是什么？

____

6. 去骨刀的用途是剔骨。

____

7. 你可以削个苹果给我吗？

____

8. 一套刀具包含了几种不同用途的刀？

____

9. 请用抹刀在蛋糕上抹上一层霜糖。

____

10. 我可以在这个搅拌碗里拌蛋黄酱和金枪鱼吗？

____

Reading

Colander

A colander is a bowl-shaped kitchen utensil with holes in it used for draining food such as pasta or rice.The perforated nature of the colander liquid to drain through while retaining the solids inside. It is sometimes also called a pastastrainer or kitchen sieve. Conventionally, colanders are made of a light metal, such as aluminium or thinly rolled stainless steel. Colanders are also made of plastic, silicone, ceramic, and enamelware. The word colander comes from the Latin *colum* meaning *sieve*.

In kitchen utensils, a *spatula* is any utensil fitting the above description. One variety is alternately named turner, and is used to lift and turn food items during cooking, such as pancakes and fillets. These are usually made of plastic, with a wooden or plastic handle to insulate them from heat.A frosting spatula is also known as palette knife and is usually made of metal or plastic.Bowl and plate scrapers are sometimes called spatulas.

British English usage

In British English a spatula is similar in shape to a palette knife without holes in the blade. A wide-bladed utensil with long holes in the blade used for turning food is a fish slice.

A frying pan, frypan, or skillet is a flat-bottomed pan used for frying, searing, and browning foods. It is typically 200 to 300 mm (8 to 12 in) in diameter with relatively low sides that flare outwards, a long handle, and no lid. Larger pans may have a small grab handle opposite the main handle. A pan of similar dimensions, but with vertical sides and often with a lid, is called a sauté pan or sauté. While a sauté pan can be used like a frying pan, it is designed for lower heat cooking methods, namely sautéing.

A coating is sometimes applied to the surface of the pan to make it non-stick. Frying pans made from bare cast iron or carbon steel can also gain non-stick properties through seasoning and use.

For some cooking preparations a non-stick frying pan is inappropriate, especially for deglazing, where the residue of browning is to be incorporated in a later step such as a pan sauce. Since little or no residue can stick to the surface, the sauce will fail for lack of its primary flavoring agent.

Non-stick frying pans featuring teflon coatings must *never* be heated above about 240 °C (464 °F), a temperature that easily can be reached in minutes. At higher temperatures non-stick coatings decompose and give off toxic fumes.

Electric frying pans

An electric frying pan or electric skillet incorporates an electric heating element into the frying pan itself and so can function independently off of a cooking stove. Accordingly, it has heat-insulated legs for standing on a countertop. (The legs usually attach to handles.) Electric frying pans are common in shapes that are unusual for "unpowered" frying pans, notably square and rectangular. Most are designed with straighter sides than their stovetop cousins and include a lid. In this way they are a cross between a frying pan and a sauté pan.

A modern electric skillet has an additional advantage over the stovetop version: heat regulation. The detachable power cord/unit incorporates a thermostatic control for maintaining the desired temperature.

With the perfection of the thermostatic control, the electric skillet became a popular kitchen appliance. Although it largely has been supplanted by the microwave oven, it is still in use in many kitchens.

Microwave Oven

A microwave oven, often colloquially shortened to microwave, is a kitchen appliance that heats food by bombarding it with electromagnetic radiation in the microwave spectrum causing polarized molecules in the food to rotate and build up thermal energy in a process known as dielectric heating. Microwave ovens heat foods quickly and efficiently because excitation is fairly uniform in the outer 25~38 mm of a dense (high water content) food item; food is more evenly heated throughout (except in thick, dense objects) than generally occurs in other cooking techniques.

Percy Spencer invented the first microwave oven after World War II from radar technology developed during the war.

Microwave ovens are popular for reheating previously cooked foods and cooking vegetables. They are also useful for rapid heating of otherwise slowly prepared cooking items, such as hot butter, fats, and chocolate.

Unit 12

Kitchen Dining Utensils and Etiquette

Lead in

Dining utensils are tools used for eating (c.f. the more general category of tableware). Some utensils are both kitchen utensils and eating utensils. Cutlery (i.e. knives and other cutting implements) can be used for both food preparation in a kitchen and as eating utensils when dining. Other cutlery such as forks and spoons are both kitchen and eating utensils.

Vocabulary

fork [fɔːk] *n.&v.* 叉子；使……成叉形；用叉叉起
skewer ['skjuːə(r)] *n.&v.* 串肉扦；用扦串肉
shish kebab ['ʃɪʃkɪbæb] *n.* 羊肉串
ladle ['leɪdl] *n.&v.* 勺子
dish [dɪʃ] *n.* 盘；碟；菜肴；一道菜
plate [pleɪt] *n.* 板；盘子；板块
bowl [bəʊl] *n.* 碗
chopstick ['tʃɑp.stɪk] *n.* 筷子
tableware [ˈteɪblweə(r)] *n.* 餐具
host [həʊst] *n.* 男主人
hostess [ˈhəʊstəs] *n.* 女主人
etiquette [ˈetɪket] *n.* 礼节

slotted spoon 有孔汤匙
soup tureen 盛汤盖碗
main course 主菜
table manner 餐桌文化

Focus on Language

Dialogue 1

Hostess: Would you like to hxdye some more chicken ?

Guest: No, thank you. The chicken is very delicious, but I'm just too full.

Host: But I hope you have some dessert. Mary makes very good pumpkin pies.

Guest: That sounds very tempting. But I hope we can wait a little while, if you don't mind.

Host: Of course. How about some coffee or tea now?

Guest: Tea, please. Thanks.

Dialogue 2

A: I used to have headache with Chinese chopsticks, but now I'm getting used to it.

B: You must have taken a lot of efforts to practice. By the way, in China, don't stick chopsticks vertically into food, especially in a bowl of rice.

A: I'm not very clear about this. Why?

B: Because chopsticks in bowl of rice usually appear in Chinese traditional memorial for ancestors.

◇◇

Task 1 Look at the pictures below and try to write down English name below the pictures.

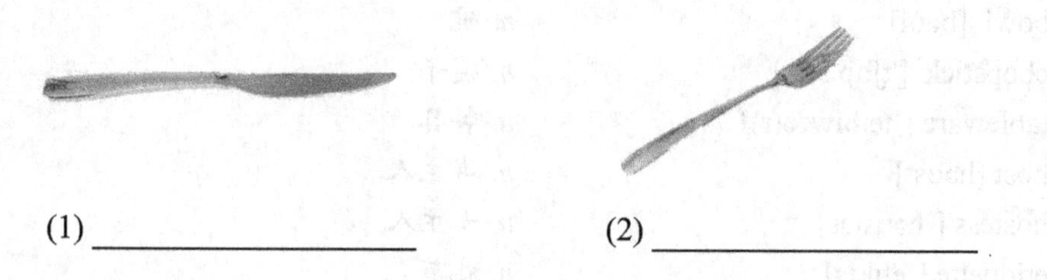

(1) ____________________ (2) ____________________

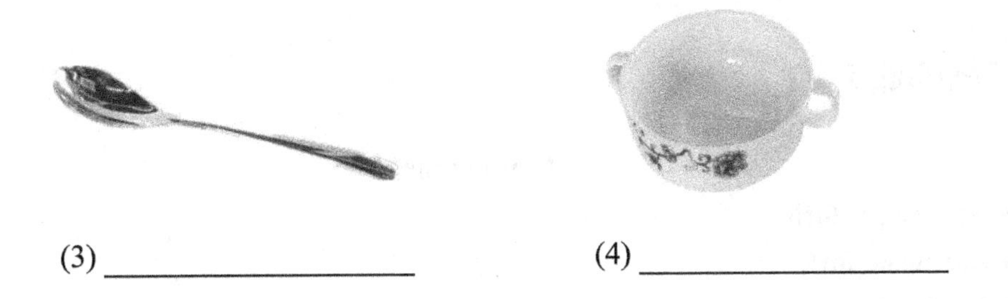

(3) ____________________ (4) ____________________

Task 2 Look at the picture below, discuss with your partner about seats arranging.

Task 3 Translate the following sentences into English.

1. 西方人喜欢用叉子，中国人喜欢用筷子。

____

2. 把鸡肉放到碗里。

____

3. 用叉子吃牛排。

____

4. 请给他一个勺子，他要喝汤了。

____

5. 用勺子挖哈密瓜。

____

6. 在用餐的时候，男女主人一般不坐在一起，便于照顾到宾客。

____

7. 客人在用餐中离开餐桌是很不礼貌的。

____

Reading 1

Types of fork

- **Asparagus fork**
- **Barbecue fork**
- **Beef fork**

A fork used for picking up meat.

- **Carving fork**

A two-pronged fork used to hold meat steady while it is being carved.

- **Cheese fork**
- **Chip fork**

A two-pronged disposable fork, usually made out of sterile wood (though increasingly of plastic), specifically designed for the eating of chips (known as french fries in North America), fried fish and other takeaway foods.

- **Cocktail fork**

A small fork resembling a trident, used for spearing cocktail garnishes such as olives.

- **Cold meat fork**
- **Crab fork**

A short, sharp and narrow three-pronged or two-pronged fork designed to easily extract meat when consuming cooked crab.

- **Dessert fork (alternatively, pudding fork/cake fork in Great Britain)**

Any of several different special types of forks designed to eat desserts, such as a pastry fork.

- **Dinner fork**
- **Fish fork**
- **Fondue fork**

A narrow fork, usually having two tines, long shaft and an insulating handle, typically of wood, for dipping bread into a pot containing sauce.

- **Fruit salad fork**

A fork used which is used to pick up pieces of fruit such as grapes, strawberries, melon and other varies types of fruit.

- **Granny Fork**
- **Ice cream fork**

A spoon with flat tines used for some deserts. See spork.

- **Knork**
- **Meat fork**
- **Olive fork**
- **Oyster fork**
- **Pastry fork**
- **Pickle fork**

A long handled fork used for extracting pickles from a jar, or an alternative name for a ball joint separator tool used to unseat a ball joint.

- **Pie fork**
- **Relish fork**
- **Salad fork**

Similar to a regular fork, but may be shorter, or have one of the outer tines shaped differently.

- **Sardine fork**
- **Sporf**

A utensil combining characteristics of a spoon, a fork and a knife.

- **Spork**

A utensil combining characteristics of a spoon and a fork.

- **Sucket fork**

A utensil with tines at one end of the stem and a spoon at the other. It was used to eat food that would otherwise be messy to eat such as items preserved in syrup.

- **Tea fork**
- **Terrapin fork**

A spoon with flat tines used for some soups. See spork.

- **Toasting fork**

A fork, usually having two tines, very long metal shaft and sometimes an insulating handle, for toasting food over coals or an open flame.

- **Novelty forks**

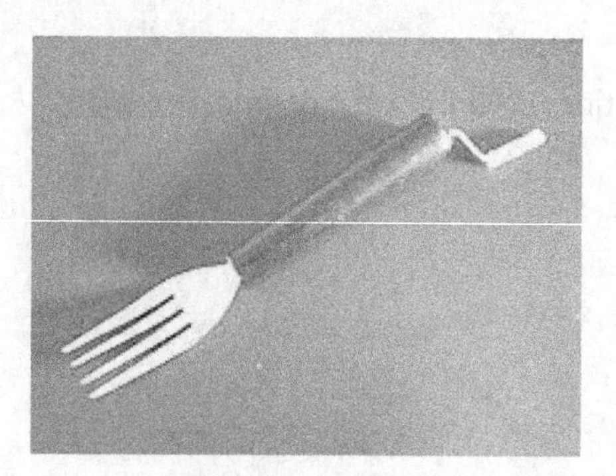

Spaghetti fork

- **Extension Fork**

A long-tined fork with a telescopic handle, allowing for its extension or contraction.

- **Spaghetti fork**

A fork with a metal shaft loosely fitted inside a hollow plastic handle.

Reading 2

Table manners

All the food a person intends to eat is put on a large plate in front of him at the beginning of the meal(分食制); if one finishes the food on his plate and wants some more, he will get a second helping.

Some families will say grace before eating. It is common with Christian families but not all families.

When the waiter serves one dish to the guests one by one, if he stands on your left, it is your turn to take the food. If on your right, it is not your turn.

As to the seating, there are places of honor. And the honored guests sit on the right side of the host and hostess. And the host and hostess often sit at the two ends of the ordinary rectangular tables. People of the same sex avoid sitting together.

Hold the knife on your right hand and fork left. After cutting the food, put your knife on the table and eat the food with your fork on your right hand.

Slurping(啧啧) and burping(打嗝) are considered as extremely rude. If you have to, say "excuse me".